Dire Predictions

UNDERSTANDING GLOBAL WARMING

Dire Predictions
UNDERSTANDING GLOBAL WARMING

Michael E. Mann and Lee R. Kump

LONDON, NEW YORK, MELBOURNE,
MUNICH AND DELHI

Book Design Richard Czapnik
Senior Editor Steve Setford
Cartographer Ed Merritt
Production Manager Silvia La Greca Bertacchi
DTP Designer David McDonald

Original Book Design Stuart Jackman
Publisher Sophie Mitchell

PEARSON
Education

Publisher Daniel Kaveney
Development Editor Erin Mulligan

First American Edition, 2008.

Published in the United States by
DK Publishing, Inc.
375 Hudson Street
New York, New York 10014

08 09 10 11 12 13 14 15 16 17 9 8 7 6 5 4 3 2 1

A catalog record for this book is available from the Library of Congress

ISBN 978-0-7566-3995-2

The papers used for the pages and the cover are FSC certified, and come from N.America, where the book was printed.

The inks used throughout are vegetable inks and the special finish on the cover is biodegradable.

OUR COMPANY is part of Pearson, a founder signatory to the UN Global Compact. This sets out a series of principles against which we measure ourselves in the areas of human rights, labor standards, the environment, and anti-corruption

High-resolution workflow proofed by Media Development and Printing Ltd, UK

Printed and bound by RR Donnelley, USA

Discover more at
www.dk.com

Contents

Part 1
CLIMATE CHANGE BASICS

Part 2
CLIMATE CHANGE PROJECTIONS

Part 3
THE IMPACTS OF CLIMATE CHANGE

Part 4
VULNERABILITY AND ADAPTATION TO CLIMATE CHANGE

Part 5
SOLVING GLOBAL WARMING

Introduction

The Intergovernmental Panel on Climate Change (IPCC) was established in 1988 to evaluate the risk of human-caused climate change. Since its inception, the IPCC's periodic assessment reports have become the *de facto* conservative standard for accuracy about the scientific facts of global climate change. Unfortunately, these assessment reports, relied upon for their accuracy and often quoted by the media and scientists alike, contain high-level scientific content that can make them difficult for the general public to understand.

As the public furor over the state of Earth's climate continues to brew, it is more important than ever for informed citizens to build a basic understanding of the reasons most scientists think the global climate is in a state of crisis. However, until now, it has been difficult for interested lay readers to find reliable sources of information that are both authoritative and easy to understand.

In this book, esteemed climate scientists Michael E. Mann (who, along with other IPCC report authors, was awarded the Nobel Prize in 2007) and Lee R. Kump have partnered with the "information architects" at DK Publishing to produce *Dire Predictions*—essential reading for citizens of a world in distress. *Dire Predictions*, at just over 200 pages, presents and expands upon the findings documented in the Fourth Assessment Report of the IPCC in an illustrated, visually stunning, and undeniably powerful way for the non-scientist.

Trouble brewing
A lone lightning bolt strikes the ground beneath an isolated "supercell" thunderstorm at sunset.

About the IPCC

The Intergovernmental Panel on Climate Change (IPCC) was established in 1988 by the United Nations Environment Program (UNEP) and the World Meteorological Organization (WMO). The Panel was tasked with preparing a scientifically based report on all relevant aspects of climate change and its impacts, and formulating possible strategies for addressing these impacts. The self-described role of the IPCC is to "assess on a comprehensive, objective, open and transparent basis the scientific, technical and socio-economic information relevant to understanding the scientific basis of risk of human-induced climate change, its potential impacts and options for adaptation and mitigation." The IPCC strives to be policy-relevant but not policy-prescriptive.

Since its inception, the IPCC has reviewed and assessed the most recent scientific, technical, and socioeconomic information on climate change at regular intervals, periodically producing a set of comprehensive, well-documented reports. The IPCC reports summarize our continually improving knowledge of the underlying science of climate and convey the most reliable available projections for future climate change and its impacts. The reports are written by thousands of the world's leading scientists. Rigorous peer review is a hallmark of the IPCC process, and expert reviewers are called upon to comment on all aspects of the reports.

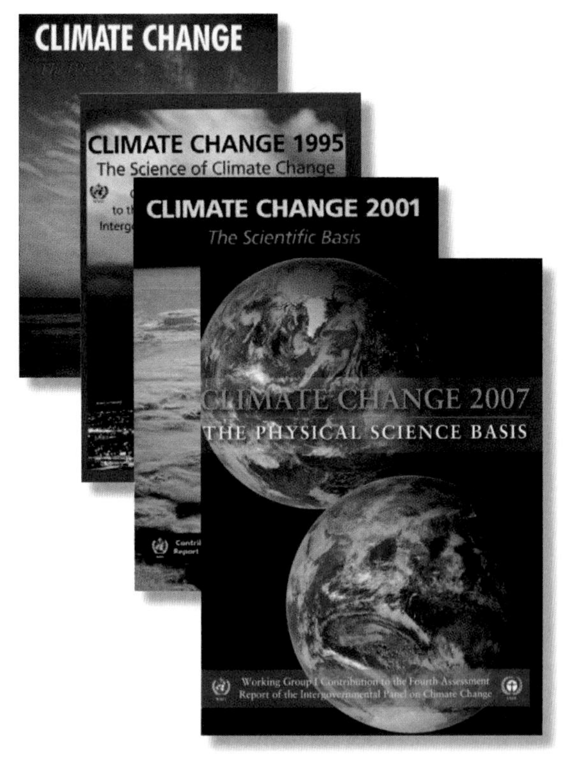

IPCC REPORTS

The information in *Dire Predictions* closely follows the findings of the IPCC Fourth Assessment Report. The authors have presented this material in a way that makes it accessible to non-scientists, and have supplemented the assessment's findings with additional and updated material.

About the authors

Dr. Michael E. Mann is a member of the Pennsylvania State University faculty, holding joint positions in the Departments of Meteorology and Geosciences, and the Earth and Environmental Systems Institute (EESI). He is also director of the Penn State Earth System Science Center (ESSC).

Dr. Mann received his undergraduate degrees in Physics and Applied Math from the University of California at Berkeley, an M.S. degree in Physics from Yale University, and a Ph.D. in Geology & Geophysics from Yale University. His research focuses on the application of statistical techniques to understanding climate variability and climate change from both empirical and climate model-based perspectives.

Dr. Mann was a Lead Author on the "Observed Climate Variability and Change" chapter of the Intergovernmental Panel on Climate Change (IPCC) Third Scientific Assessment Report published in 2001, and a reviewer for the most recent Fourth Report. He has been organizing committee chairperson for the National Academy of Sciences "Frontiers of Science" and has served as a committee member and an advisor for other National Academy of Sciences panels. Dr. Mann is the recipient of several fellowships and prizes, and has been named to the "Scientific American 50," a list of leading visionaries in science and technology. He is author of more than 100 peer-reviewed and edited publications, and is a co-founder of RealClimate.org which seeks to inform the public, journalists, and policy makers about the science of climate change.

Dr. Lee R. Kump is a Professor in the Department of Geosciences at Pennsylvania State University, and an associate of the Penn State Earth and Environmental Systems Institute, Earth System Science Center, and the Penn State Astrobiology Research Center.

Dr. Kump received his bachelor's degree in geophysical sciences from the University of Chicago and his Ph.D. in Marine Science from the University of South Florida. He is a fellow of the Geological Society of America and the Geological Society of London. In his research he uses a variety of tools, including geochemical analysis and computer modeling, to investigate climate and biospheric change during periods of extreme and abrupt environmental and biodiversity change in Earth's history.

Dr. Kump is an active researcher with over 75 peer-reviewed and edited publications. His research has been featured in documentaries produced by *National Geographic*, the British Broadcasting Corporation (BBC), *NOVA Science-Now*, and the Australian Broadcasting Corporation. He is the lead author on the preeminent textbook in Earth System Science, *The Earth System*, now in its second edition. He is on the Board of Reviewing Editors for Science as well as the editorial board of the journal *Geobiology*. He is the associate director of the Earth System Evolution Program of the Canadian Institute for Advanced Research. He also currently serves on the National Academy of Science committee for evaluating "The Importance of Deep-Time Geologic Records for Understanding Climate Change Impacts."

What's up with the weather (and the climate!)?

You have no doubt heard quite a bit over the past few years about climate change and global warming. To truly understand these terms, and to appreciate how and why human activity is causing Earth's climate to change, you need first to understand what climate is; how it differs from weather; what factors affect it; and how modern human activity is altering it. The purpose of this section of the book is to introduce you to these concepts.

Climate and weather and us

We plan our daily activities around the weather. Will it rain? Is a storm or a cold front approaching? Weather is highly variable, and, although considerable improvements in weather forecasting have been made, largely unpredictable. Climate, on the other hand, varies more slowly and is highly predictable. We know what to expect of our local climate and the climate of many familiar regions. Panama, for example, is persistently warm and wet. Residents of Siberia and northern Alaska expect long and bitterly cold winters. In the mid-latitudes, a summer day is almost certainly going to be warmer than a winter day. Climate represents the average of many years' worth of weather. This averaging process smoothes out the individual blips caused by droughts and floods, tornadoes and hurricanes, and blizzards and downpours, while emphasizing the more typical patterns of temperature highs and lows and precipitation amounts.

The reason that climate is so predictable is that it is dependent on relatively fixed features of Earth. These include Earth's spherical form, the shape of its orbit around the Sun, and its tilted axis of rotation. Other factors that determine climate have to do with the fact that Earth possesses both oceans and continents and a multi-layered atmosphere that is composed of various gases including, critically, the greenhouse gases (the importance of which we will explain in the following pages).

Climate and latitude

Radiation from the Sun plays a big role in Earth's climate. The amount of radiation Earth receives from the Sun depends fundamentally on latitude. At the equator, the Sun's rays are most directly overhead and most directly focused on Earth's surface. As we move poleward, the Sun's position at noon is lower in the sky, and so its energy is spread over a larger area, making it less intense. This is the fundamental reason

In the far north energy from the Sun is dispersed.

In the tropics energy from the Sun is concentrated.

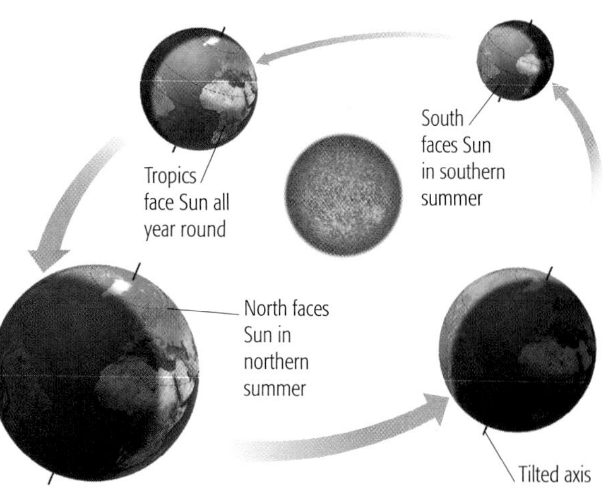

that the tropics (the region between the tropic of Cancer at $23\frac{1}{2}$ °N and the tropic of Capricorn at $23\frac{1}{2}$ °S) are warm and the poles are so cold.

Another factor that changes with latitude is seasonal contrast: how hot the summers are, and how cold the winters are. In the tropics the difference in temperatures between summer and winter is fairly subtle, whereas at mid- to high-latitudes, the difference is quite sizable. However, the existence of the seasons themselves depends not on latitude *per se*, but on the fact that Earth's spin axis, the imaginary line that runs from pole to pole through the center of Earth, is tilted. Summer occurs in either hemisphere when the spin axis is inclined toward the Sun, while winter occurs when it is tilted away. The impact of spin axis tilt is most pronounced above the Arctic and Antarctic circles. This is why in these regions the Sun shines 24 hours a day during the summer and they remain in perpetual darkness during the winter.

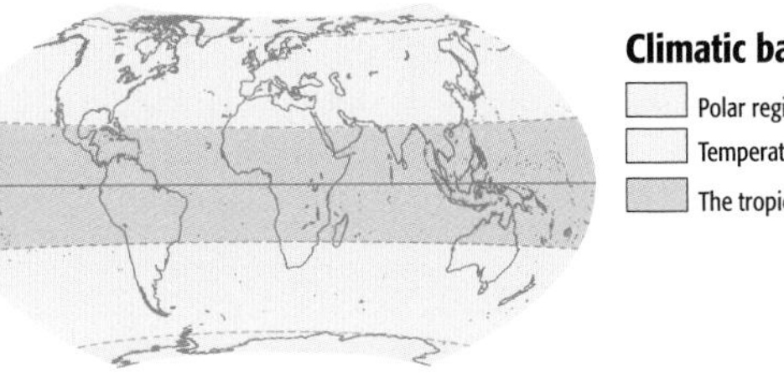

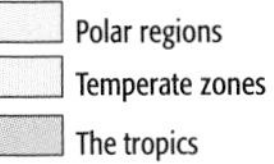

Climate and the oceans

Another important factor determining continental climate is proximity to oceans. Water has a tremendous capacity for storing heat, much greater than that of the land. The oceans warm slowly during the summer and cool slowly during the winter, so coastal regions benefit from their moderating influence. In contrast, the continental interiors respond quickly to seasonal changes. This is why places like North Dakota and Saskatchewan typically have warm summers and cold winters compared to coastal locations.

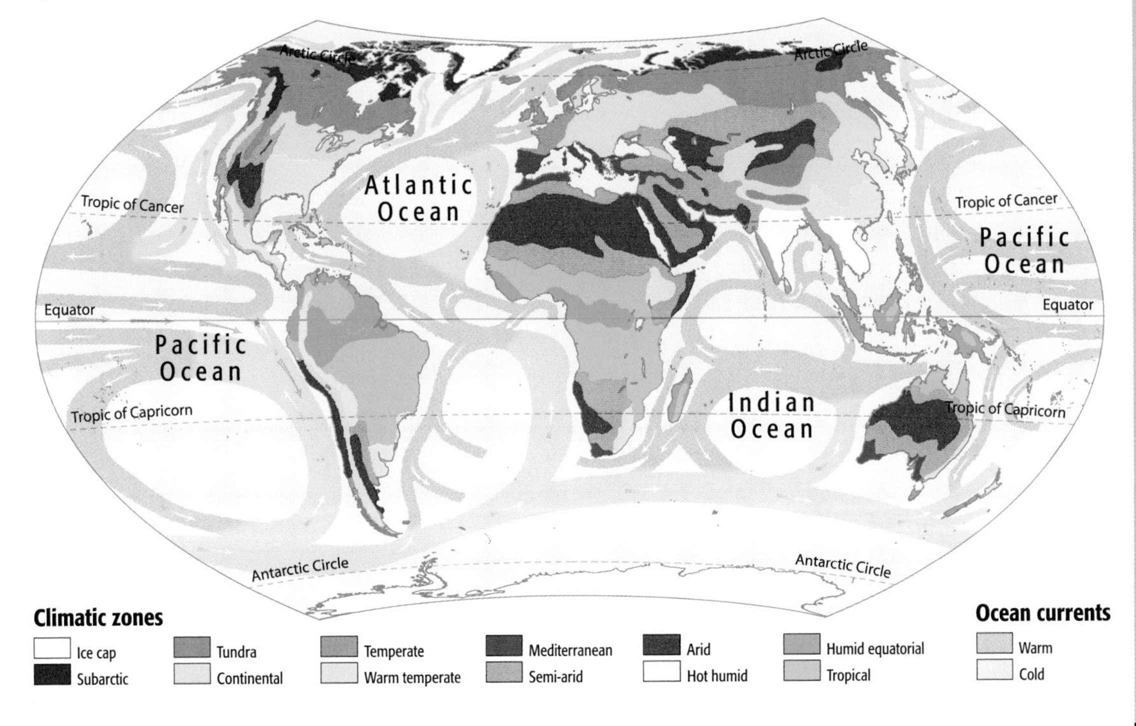

Climate and the atmosphere

The existence and composition of Earth's atmosphere also influences the climate. The atmosphere is the gaseous envelope surrounding Earth; it is retained by Earth's gravitational pull. Our atmosphere features distinct layers. The first 64–80 km above the surface contains 99% of the total mass of the atmosphere and is generally uniform in gas composition with some notable exceptions, including large variations in water vapor and high concentrations of ozone, known as the ozone layer, at 19 to 50 km.

Atmospheric composition

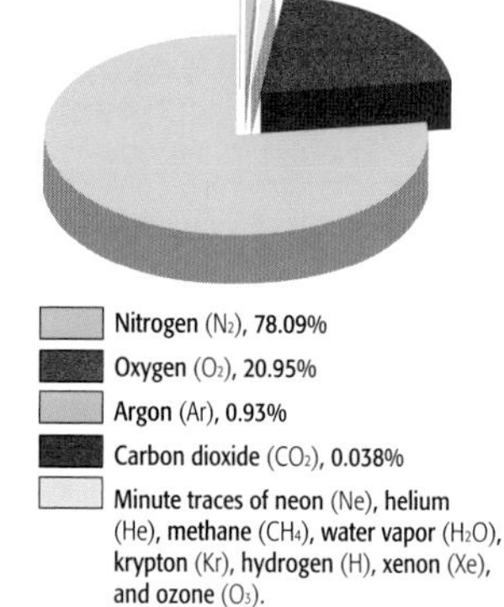

- Nitrogen (N_2), 78.09%
- Oxygen (O_2), 20.95%
- Argon (Ar), 0.93%
- Carbon dioxide (CO_2), 0.038%
- Minute traces of neon (Ne), helium (He), methane (CH_4), water vapor (H_2O), krypton (Kr), hydrogen (H), xenon (Xe), and ozone (O_3).

Exosphere
Outermost layer of the atmosphere. Extends to about 10,000 km.

Thermosphere
Extends to about 640 km.

Mesosphere
The portion of the atmosphere from about 50 to 80 km above the surface. Air becomes cooler as the altitude increases.

Stratosphere
Extends upward to a height of about 50 km. Contains atmospheric ozone layer. Temperature increases with altitude through the stratosphere, inhibiting vertical air currents, and making the stratosphere highly stable, in contrast to the troposphere.

Troposphere
Layer in contact with Earth's surface. Extends upward from the surface to about 8 km to 17 km. Air temperature decreases with altitude leading to instability. Less dense air sits below more dense air, which results in air movements and storm generation. "Weather" takes place almost exclusively within the troposphere.

Sea level

Mesopause
Boundary between the mesosphere and the thermosphere.

Stratopause
Boundary between the stratosphere and the mesosphere.

Atmospheric ozone layer
Layer within the stratosphere. Absorbs ultraviolet solar radiation so warming the surrounding atmosphere.

Tropopause
Boundary between the troposphere and the stratosphere.

OZONE LAYER

Radiation can be beneficial. Our planet would be cold without the Sun's rays, and plants need solar radiation to photosynthesize. But radiation can also be dangerous, particularly ultraviolet (UV) radiation. Fortunately, oxygen and ozone molecules in the stratosphere absorb most of the UV radiation reaching Earth. Ozone is a compound of oxygen that contains three atoms (the oxygen gas we breathe contains two oxygen atoms) and it is a lung irritant and smog producer when encountered in surface air pollution. However in the stratosphere, ozone protects life on Earth by absorbing UV radiation. In the process, the radiation destroys the chemical bonds between the oxygen atoms in the ozone molecule. Under normal circumstances, the ozone molecules rapidly reform. Unfortunately by adding chlorofluorocarbons to the atmosphere, which, like radiation, destroy ozone molecules, we've unwittingly accelerated the destruction of the ozone layer, reducing its effectiveness in protecting us from UV radiation.

Atmospheric circulation

To understand rainfall patterns, a major player in climate, we need to understand the basic principles of atmospheric circulation.

The pattern of rising moist air near the equator and sinking dry air in the subtropics is referred to as the "Hadley Circulation." The Hadley Circulation is a key component of the general circulation of the atmosphere; it helps to transport heat from the equatorial region to higher latitudes. Because of the Hadley Circulation, generally the tropics are warm and wet, while the subtropics are warm and dry. And as a result of the atmospheric circulation patterns found at higher latitudes, the mid-latitude regions experience large seasonal contrasts in temperature and rainfall patterns, while the polar regions are generally cold and dry. Rainfall in the mid-latitudes is related to the "polar front." Those of us who live in North America or Europe may know the polar front by a different name—the "storm track"—an expression that refers to the day-to-day variations in the location and intensity of the polar front.

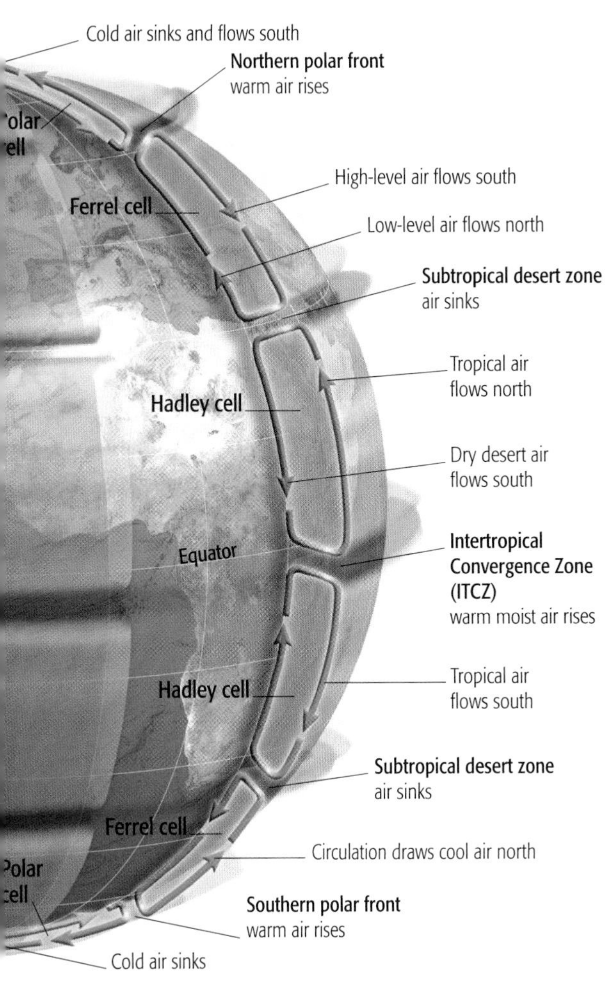

Basic principles of atmospheric circulation

1. Water evaporates from the land and the ocean and becomes water vapor, a gas that composes part of the lower atmosphere.
2. Like a huge hot-air balloon, air near the ground in the tropics warms as a result of solar radiation, becomes buoyant, and rises.
3. As the warm tropical air rises it expands, and, like gas coming out of a spray can, it cools.
4. Cold air can hold less water vapor, so as the rising tropical air cools, water condenses out as droplets that congeal and form towering cumulus clouds and rainfall-producing thunderstorms. Thus, the tropics are rainy.
5. The rising air in the tropics draws air in from higher latitudes, forming the Intertropical Convergence Zone (ITCZ). The ITCZ migrates north and south within the tropics as the seasons change. The ITCZ is associated with trade winds that converge near the equator.
6. Air rising in the tropics moves poleward once it reaches higher altitudes. Because Earth is spinning, this poleward flow gets disrupted, and air sinks at approximately 30°S and 30°N (the subtropics).
7. This air sinking in the subtropics is now quite dry because most of the water vapor was already precipitated out of it when the air was rising. Furthermore, as air descends it gets compressed, and, like an inflating bicycle tire, warms, which dries it out even more. This is why deserts tend to occur at these subtropical latitudes.
8. A second region of rising air exists in middle-to-high latitudes (roughly 40–60°N and 40–60°S) in the region known as the polar front. Here, warm air from lower latitudes encounters cold polar air heading towards the equator. The denser polar air forces itself underneath the warmer air mass, causing it to rise, cool, and condense out its water vapor.
9. Finally, air near the poles sinks, causing the polar regions to be arid.

GREENHOUSE GASES AND EARTH'S CLIMATE

The greenhouse effect occurs on our planet because the atmosphere (the gaseous cloud that surrounds Earth) contains greenhouse gases. Greenhouse gases are special in that they absorb heat. In doing so, they warm the atmosphere around them. When this happens, our climate changes and global warming occurs. Greenhouse gases exist naturally in Earth's atmosphere in the form of water vapor, carbon dioxide, methane, and other trace gases, but atmospheric concentrations of some greenhouse gases, such as carbon dioxide and methane, are being increased as a result of human activity. This occurs primarily as a result of the burning of fossil fuels, but also through deforestation and agricultural practices. Certain greenhouse gases, such as CFCs, and the surface ozone found in smog (which is distinct from the natural ozone found in the lower stratosphere), are produced exclusively by human activity.

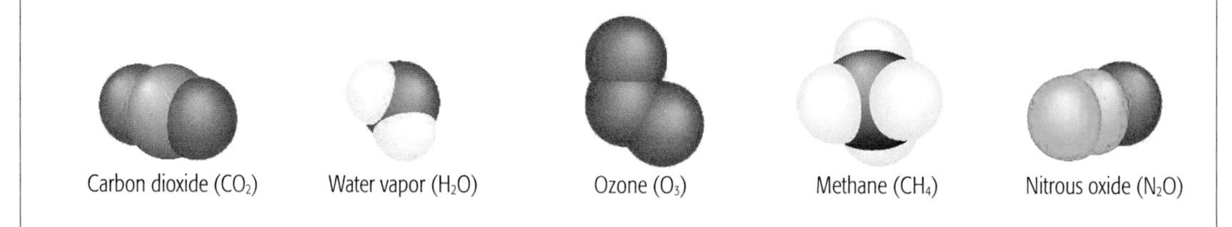

Carbon dioxide (CO_2) Water vapor (H_2O) Ozone (O_3) Methane (CH_4) Nitrous oxide (N_2O)

Climate history, climate change, the greenhouse effect, and us

Looking back on Earth's history, it comes as no surprise that climates change. Indeed, on any timescale—decadal, century, millennial, or over millions of years—the climate record is anything but constant. Over the last two million years, as ice sheets advanced and retreated across northern North America and Scandinavia, climates have oscillated between very cold and more pleasant, like today. Geologists have designated this interval of time an ice age and divide it into two epochs, the Pleistocene, which lasted from 2 million years ago until 10,000 years ago, and the Holocene, which encompasses the last 10,000 years. Ice ages are marked by episodes of extensive glaciation, alternating with episodes of relative warmth. The colder periods are called glacials, the warmer periods are referred to as interglacials. The Holocene is the most recent interglacial period of Earth's most recent ice age.

Prior to the Holocene epoch, just 20,000 years ago the world was gripped by a glacial climate with ice sheets covering much of North America and Scandinavia. Prior to 2 million years ago there were no large ice sheets in the northern hemisphere, and prior to 34 million years ago there were no large ice sheets anywhere. Throughout the time of the dinosaurs (referred to as the Mesozoic Era, which spanned from 252 to 65 million years ago) the world was considerably warmer than today, and during the warmest intervals, reptiles and other cold-intolerant organisms lived above the Arctic circle. We have to go back into "deep time," more than 300 million years ago, to find evidence for a previous ice age, one that included glacial periods perhaps considerably colder and more extensive than the most recent Pleistocene glacial epoch.

Despite this backdrop of natural climate fluctuations on various timescales, human greenhouse gas emissions (see box above), are creating an atmosphere unlike anything Earth has experienced for tens of millions of years. In many respects there may be no geologic precedent for the accelerated rate of the disturbance we are imposing on Earth's climate system; the resulting impacts may be quite unlike those associated with past natural climate variation.

So how do we, as individual citizens, best address this problem? Becoming informed about the nature of the threat, and the potential solutions that are available, is a key first step. Hopefully this book, above all else, will serve to equip readers with the information necessary to make wise choices—because it is becoming increasingly clear that the decisions we make impact our collective future world.

ICE KINGDOMS

Scientists refer to the cold regions of the planet where water persists in its frozen form (i.e., regions covered with glaciers and ice sheets or with permanently frozen soils) as the cryosphere. Much of the cryosphere exists near the poles, but high-altitude mountain glaciers occur at lower latitudes. Glaciers are huge masses of ice formed from compacted snow. An ice sheet is a mass of glacier ice that covers surrounding terrain and is greater than 50,000 square kilometers. The only current ice sheets are in Antarctica and Greenland.

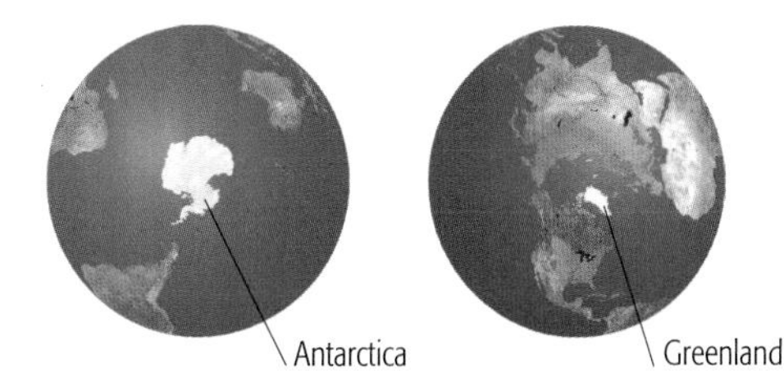

The two most important regions of the cryosphere are the continental ice sheets of Antarctica and Greenland. These huge expanses of glacial ice significantly affect the amount of solar energy reflected to space, but their most significant role is their storage of water. If the ice sheets were to melt completely, sea level would rise by about 80 m. Much of this storage is in the East Antarctic ice sheet, which is less likely to be affected by anthropogenic warming in the next few centuries; West Antarctica and Greenland melting would cause a more modest but nevertheless devastating 11 m of sea-level rise. In contrast, the expansion and contraction of sea-ice (floating ice near the poles) has no effect on sea level, but can dramatically affect ocean circulation, local climate, and ecosystems. Perennially frozen ground (permafrost) influences soil water content and vegetation over vast regions and is one of the cryosphere components most sensitive to atmospheric warming trends (see p.138). Other regions of the cryosphere are also responding to climate change: the seasonal minimum sea-ice coverage of the Arctic Ocean is currently diminishing (see p.98), and most mountain glaciers are shrinking (see p.59).

Aerial view of the edge of the Greenland ice sheet
The Greenland and Antarctic ice sheets have largely survived the glacial/interval fluctuations of the last 2 million years, whereas the North American (Laurentide) and Scandinavian (Fennoscandian) ice sheets have come and gone.

Part 1 Climate Change Basics

Basic principles of physics and chemistry dictate that Earth will warm as concentrations of atmospheric greenhouse gases increase. Though various natural factors can influence Earth's climate, only the increase in greenhouse gas concentrations linked to human activity, principally the burning of fossil fuels, can explain recent patterns of global warming. Other changes in Earth's climate, such as shifting precipitation patterns, worsening drought in many locations, increasingly severe heat waves, and more intense Atlantic hurricanes, are also likely repercussions of human's impact.

The relative impacts of humans and nature on climate

A variety of human actions as well as natural factors can potentially affect Earth's climate.

Natural impacts

Natural factors influencing climate include:

- The Sun. Over time, small but measurable changes occur in the output of the Sun—Earth's ultimate source of warming energy.
- Volcanic eruptions. Explosive volcanic eruptions modify the composition of the atmosphere by injecting small particles called "aerosols" into the atmospheric layer known as the stratosphere (see p.12), where they may reside for several years. These particles either reflect or absorb incoming solar radiation that would otherwise warm Earth's surface.
- Earth's orbit. While changes in Earth's orbit relative to the Sun influence climate on timescales of many millennia, they are not thought to play any significant role on the shorter timescales relevant to modern climate change.

Mount Pinatubo
The 1991 Mount Pinatubo eruption in the Philippines was the most explosive volcanic eruption of the 20th century. It had a cooling effect on Earth's surface for several years after the eruption.

Human impacts

The main human impact on climate is an enhanced greenhouse effect (see p.22), leading to a warming of the lower atmosphere. This is caused by increases in the atmospheric concentrations of greenhouse gases, primarily carbon dioxide produced by fossil-fuel burning (see p.26). There are also several secondary impacts of human activity. One of these involves the introduction into the atmosphere of aerosols like the ones ejected by volcanoes. These small particles (mostly sulfate and nitrate) are suspended in the atmosphere by industrial activity, such as coal combustion. Industrial aerosols reside in the lower atmosphere for only a short amount of time, and therefore must constantly be produced in order to have a sustained climate impact. The impacts of aerosols are more regionally limited and more variable than those of the well-mixed greenhouse gases. Aerosols generally reflect solar radiation back into space, and therefore represent a regional cooling influence overall. However, certain aerosols (including black carbonaceous aerosols) can, like greenhouse gases, have a surface warming influence instead. Other human impacts include stratospheric ozone depletion (see p.30) and changes in land use such as tropical deforestation, which modifies the absorptive and energy-exchange properties of Earth's surface.

Industrial pollutants
The smokestacks of factories such as this paper mill spew greenhouse gases in the form of carbon dioxide and nitrous oxide. They also produce significant amounts of aerosol-forming sulfur dioxide.

Why is climate changing?

Because of the different ways that these factors influence the patterns of both solar and longwave radiation reaching Earth, it is possible to distinguish which factors are most likely responsible for any given observed change in climate. Indeed, scientists have now determined that while natural factors have been responsible for substantial changes in climate in past centuries and millennia, human impacts, increased greenhouse gas concentrations in particular, appear to be responsible for the major climate changes of recent decades (see p.34).

Human impacts appear to be responsible for the major climate changes of recent decades.

Taking action in the face of uncertainty

The role of the scientist in global policy making

POSSIBLE PATHS OF FUTURE GLOBAL WARMING

There is considerable scientific uncertainty about how much global warming will occur and how fast it will happen, partly because the key socioeconomic factors that determine rates of fossil-fuel consumption are so unpredictable. For this reason, in the IPCC report, scientists are careful to refer to "projections" rather than "predictions" when discussing future emissions scenarios and their climatic and socioeconomic implications. Here we see the path warming has taken in the recent past and several possible projected outcomes, each of which corresponds to a different future fossil-fuel use scenario (see p.86). The right panel indicates the uncertainty associated with climate model projections for three scenarios.

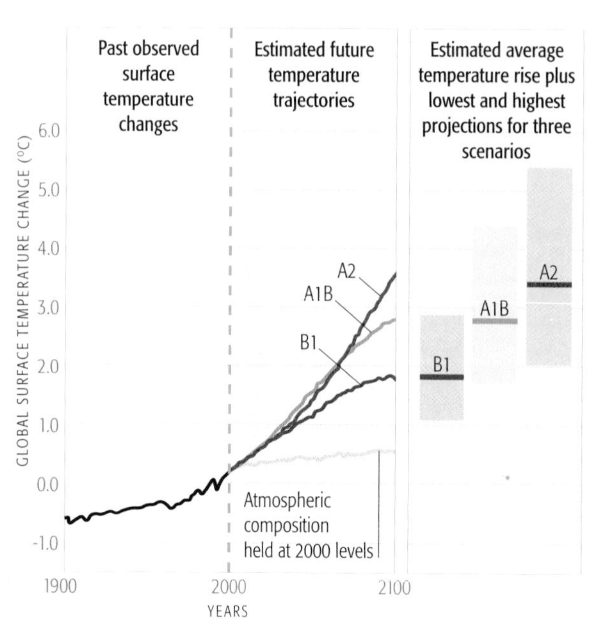

Uncertainty in global warming projections exists whether we like it or not. Some people express skepticism in response to this uncertainty, and cite it as an excuse for inaction. Scientists themselves are trained to be skeptical. They recognize that few things in science can be stated with certainty, that hypotheses can only be disproved, not proved, and that results and conclusions should be expressed in terms of this uncertainty. Unfortunately, while scientists are able to make strong conclusions from uncertain results, others view uncertainty as an indicator of flawed or inadequate scientific approaches.

Why are climate projections uncertain?

In climate science, uncertainty arises from a variety of sources, such as the inherently unpredictable nature of both the physical climate system and the human factors driving climate change, the necessary simplifications that occur when computer models are created, and incomplete knowledge about critical parameters in these models. In determining the likelihood that a conclusion is correct, climate scientists often turn to statistics, but some factors cannot be quantified by data. In these cases, likelihoods can only be established based on expert judgment. Scientific conclusions arise from time-tested theories, accurate observations, realistic models based on the fundamentals of physics and chemistry, and consensus among colleagues working in the discipline.

The Fourth Assessment Report

The Fourth Assessment Report of the IPCC presents conclusions in terms of the likelihood of particular outcomes.

These are expressed as a probability, based either on calculations or expert opinions. Likelihood ranges from virtually certain (greater than 99% probability of occurrence) to exceptionally unlikely (less than 1% probability of occurrence).

As policymakers are well aware, the risk associated with any of these projections is the combination of the probability of occurrence and the severity of the damage if it were to occur. This means that we should not ignore the projections toward the bottom of the table. For example, although it is unlikely that the Antarctic and Greenland ice sheets will collapse during the 21st century, if they were to collapse, the consequences would indeed be dire. Therefore, the risk of the ice sheets collapsing is actually quite high.

IPCC PROJECTIONS FOR THE 21ST CENTURY

This table outlines the IPCC's projections for the 21st century, ranked in decreasing order of certainty.

Projection	Probability
▪ Cold days and nights will be warmer and less frequent over most land areas ▪ Hot days and nights will be warmer and more frequent over most land areas	VIRTUALLY CERTAIN **99%**
▪ If the atmospheric CO_2 level stabilizes at double the present level, global temperatures will rise by more than 1.5°C ▪ The warming over inhabited continents by 2030 will be about double the observed variability during the 20th century ▪ There will be an observed increase in methane concentration due to human activities ▪ The rate of increase in atmospheric CO_2, methane, and nitrous oxide will reach levels unprecedented in the last 10,000 years ▪ The frequency of warm spells and heat waves will increase ▪ The frequency of heavy precipitation events will increase ▪ Precipitation amounts will increase in high latitudes ▪ The ocean's conveyor-belt circulation will weaken or shut down abruptly	VERY LIKELY **90%**
▪ If the atmospheric CO_2 level stabilizes at double the present level, global temperatures will rise by between 2°C and 4.5°C ▪ The future increase in global average surface temperature will be between −40% and +60% of the values predicted by climate models ▪ Areas affected by drought will increase ▪ The number of frost days will decrease, and growing seasons will lengthen ▪ Intense tropical cyclone activity will increase, with greater wind speeds and heavier precipitation ▪ Extreme high-sea-level events will increase, as will ocean wave heights of mid-latitude storms ▪ Precipitation amounts will decline in the subtropics ▪ The loss of glaciers will accelerate in the next few decades ▪ Climate change will promote ozone-hole expansion, despite an overall decline in ozone-destroying chemicals	LIKELY **66%**
▪ The West Antarctic ice sheet will pass the melting point if global warming exceeds 5°C	ABOUT AS LIKELY AS NOT **35–50%**
▪ Antarctic and Greenland ice sheets will collapse due to surface warming	UNLIKELY **33%**
▪ The ocean's conveyer-belt circulation will suffer an abrupt transition ▪ If the atmospheric CO_2 level stabilizes at double the present level, global temperatures will rise by less than 1.5°C	VERY UNLIKELY **10%**

0 10 20 30 40 50 60 70 80 90

PROBABILITY (%)

Why is it called the greenhouse effect?

Unfortunately, the label has stuck, but the greenhouse effect in our atmosphere is not exactly like an actual greenhouse. A greenhouse lets in solar energy (mostly in the form of visible light), which keeps it warm and allows the plants inside to grow. The greenhouse stays warm primarily because its glass windows prevent the wind from carrying away the heat. This is very different from the greenhouse effect.

The greenhouse effect occurs on our planet because the atmosphere (the gaseous cloud that surrounds Earth) contains greenhouse gases. Greenhouse gases are special in that they absorb heat. In doing so, they warm the atmosphere around them. Not all gases are greenhouse gases. In fact, nitrogen and oxygen—the most abundant gases in the atmosphere—aren't greenhouse gases. Fortunately for life on Earth,which depends on some atmospheric warming to exist, other gases are, including water vapor, carbon dioxide, and methane. Without its greenhouse atmosphere, Earth's temperature would plummet to well below freezing.

We know that Earth has been a habitable planet for over 3 billion years. This means that there has always been a greenhouse effect. The carbon dioxide that humanity is adding to the atmosphere today isn't creating the greenhouse effect, it's simply intensifying it.

Hot house
The greenhouse effect does keep the planet warm like the plants inside this greenhouse, but it functions somewhat differently than a real greenhouse.

HOW THE GREENHOUSE EFFECT WORKS

Greenhouse gases allow sunlight to pass through the atmosphere and heat Earth, but they interfere with the loss of heat from the land and ocean, redirecting some of that heat back to the surface.

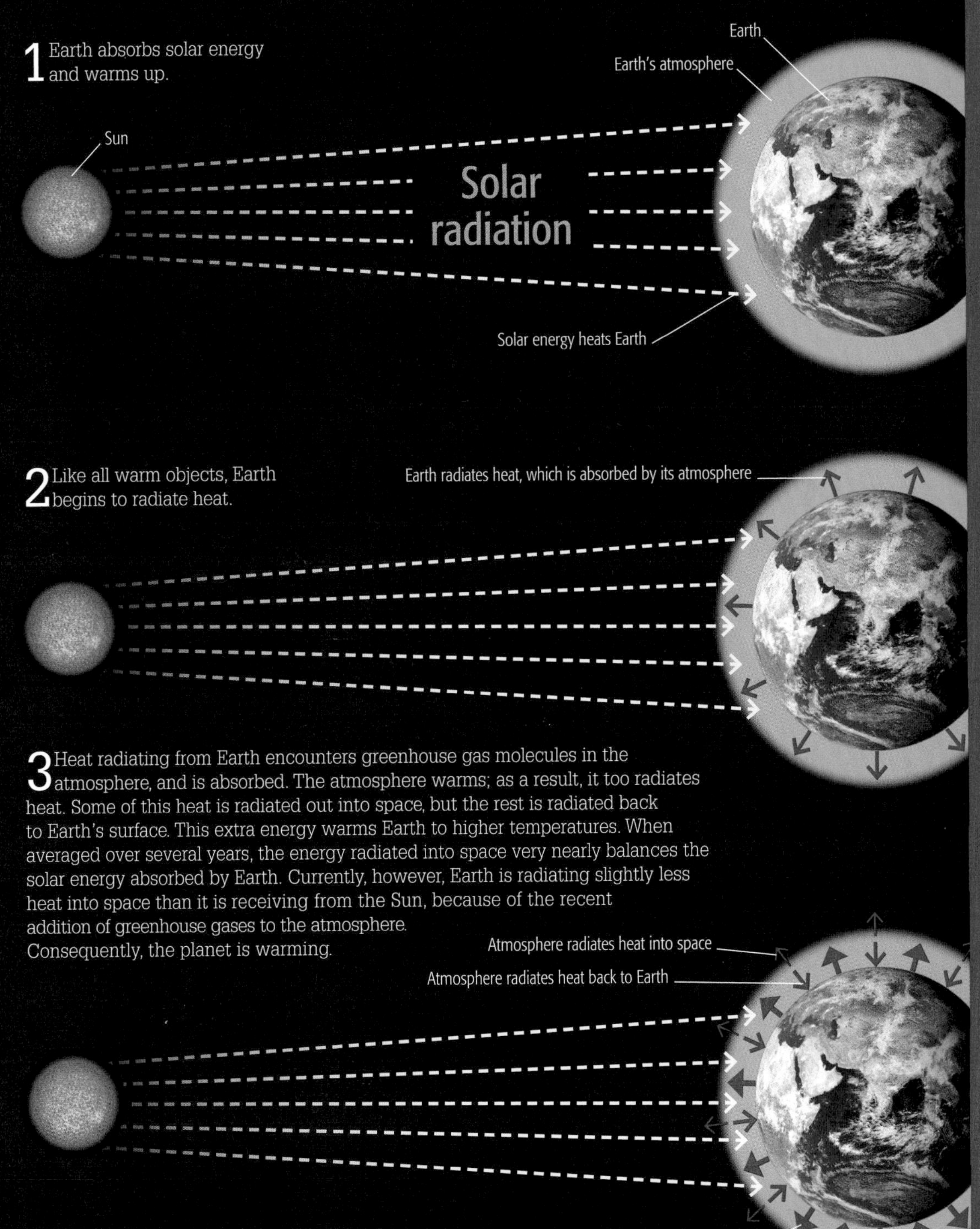

Feedback loops compound the greenhouse effect

When we think about the effects of adding carbon dioxide and other greenhouse gases to the atmosphere, we have to think not just about what these gases themselves might do to climate directly, but also about their indirect effects. This is particularly important when addressing the criticism leveled by global warming skeptics that climate researchers over-emphasize the effects of carbon dioxide (CO_2), while ignoring the fact that water vapor is in fact the most powerful of the greenhouse gases. Changes in water vapor content, however, occur only primarily in response to warming or cooling, which itself is caused by changes in atmospheric CO_2. Indirect effects, such as water vapor fluctuations, are often the result of what are known as "feedback loops."

Direct radiative effect

Climate scientists refer to the "direct radiative effect" of carbon dioxide and other greenhouse gases—that is, the effect that a particular gas has on the energy budget of the planet. In terms of direct radiative effect, carbon dioxide is important, but water vapor is an even larger contributor to the overall greenhouse effect. Knowing how much indirect warming greenhouse gas emissions will cause is trickier, because of the complex feedback loops that are set in motion when greenhouse gases are added to the atmosphere.

Positive feedback

Adding carbon dioxide to the atmosphere tends to warm the atmosphere. The

POSITIVE FEEDBACK LOOP

- Adding carbon dioxide to the atmosphere tends to warm the atmosphere, causing global warming.
- The warm atmosphere causes surface water to evaporate and become water vapor.
- Since water vapor is a greenhouse gas, the atmosphere tends to warm even more as water vapor increases.

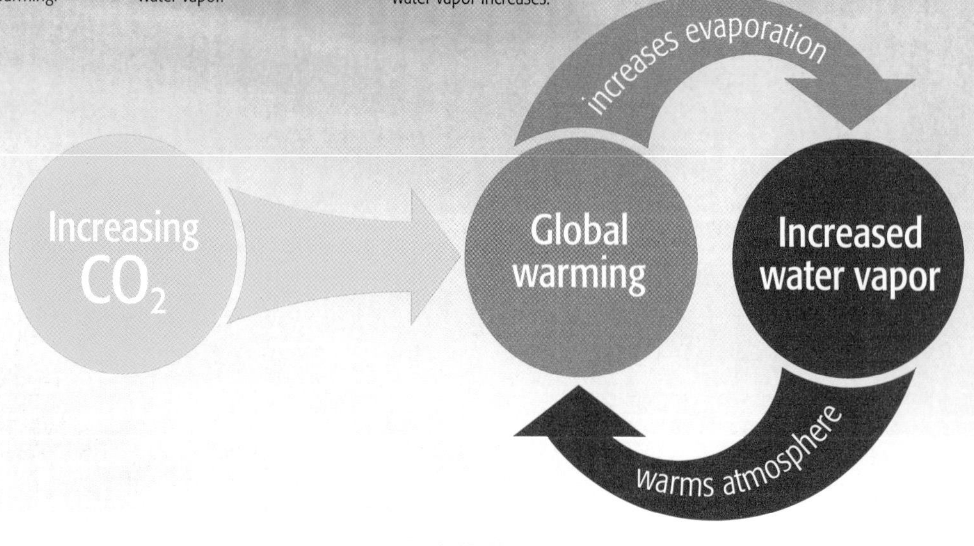

initial warming causes surface water to evaporate. Since water vapor is a greenhouse gas, the atmosphere will then tend to warm even more. This effect, known as the "water vapor feedback loop," is a positive feedback loop, because it amplifies the original change. Similarly, a modest amount of warming at high latitudes (e.g., Alaska and Scandinavia) can lead to a substantial melting of snow and ice, exposing the soil, rocks, or the ocean below. Because these surfaces are less reflective than snow, they absorb more solar radiation, thereby warming even more rapidly. This effect constitutes another very important positive feedback loop.

Negative feedback

Some of the additional water vapor in the atmosphere will condense to form clouds. Clouds contribute to the greenhouse effect by trapping heat in the atmosphere, but they also reflect solar energy back to space, helping to cool the planet. Depending on where the clouds form, their overall effect therefore may be to either cool or warm the atmosphere. So things become even more complicated. If low clouds become more prevalent in response to increased CO_2, they have a cooling effect, thus offsetting some of the initial warming. This is a negative feedback effect, as it diminishes the original change.

Observations and modeling demonstrate that the overall effect of clouds is to cool the planet; this means that the warming induced by the buildup of carbon dioxide is likely to be somewhat less than it would be if the only role that water vapor played was that of an additional greenhouse gas. Nevertheless, positive feedbacks in the climate system outweigh negative feedbacks, so the expected warming from CO_2 buildup is greater than its direct radiative effect alone.

Clouds from both sides
Clouds, such as the cirrus ones shown here, can be involved in both negative and positive climate feedback loops.

NEGATIVE FEEDBACK LOOP

- Adding carbon dioxide to the atmosphere tends to warm the atmosphere, causing global warming.
- The warm atmosphere causes surface water to evaporate and become water vapor.
- Some water vapor condenses to form clouds. Clouds contribute to the greenhouse effect by trapping heat in the atmosphere, but they also reflect solar energy back to space, helping to cool the planet.

increases evaporation
Increasing CO_2
Global warming
Increased low clouds
cools atmosphere

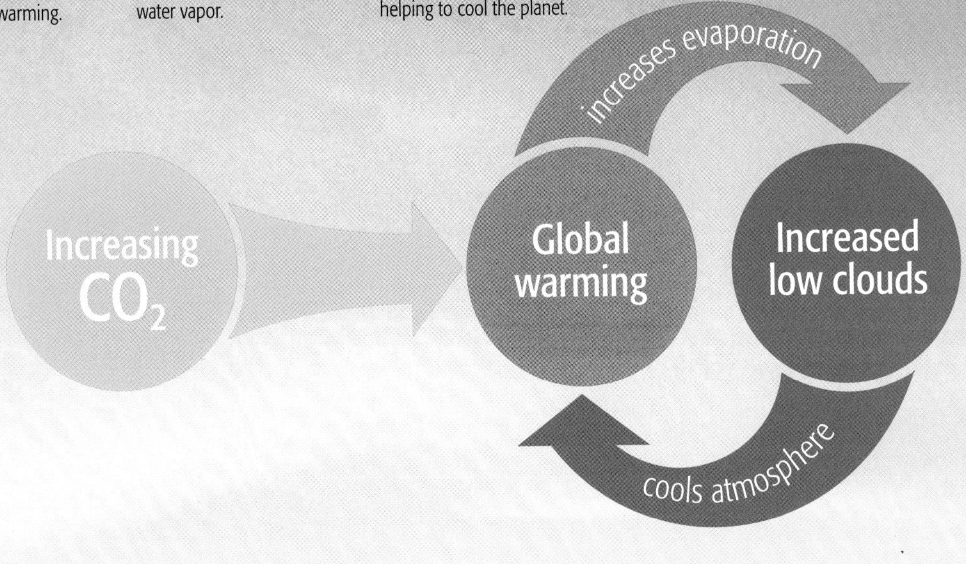

What are the important greenhouse gases, and where do they come from?

Although carbon dioxide (CO_2) has been the primary focus of concern in human-induced climate change, there are a number of other anthropogenic (human-generated) gases that also affect the radiation balance of the planet. Most of these aren't exclusively anthropogenic; with the exception of the CFCs (chlorofluorocarbons), they exist naturally. In fact, some of these gases are produced and consumed by natural processes at tremendous rates.

CARBON RECYCLING

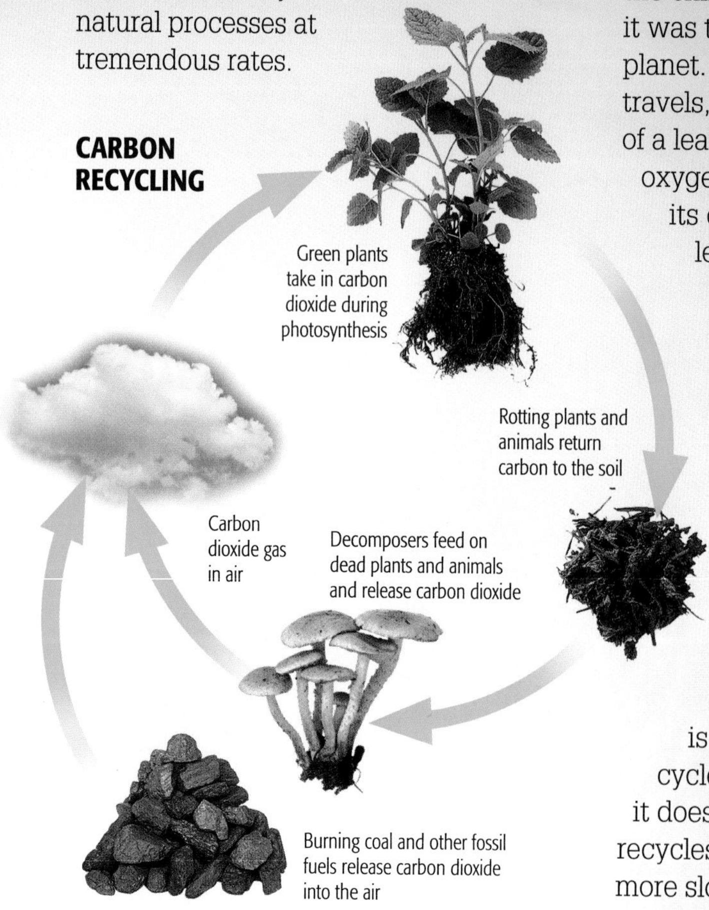

Green plants take in carbon dioxide during photosynthesis

Rotting plants and animals return carbon to the soil

Carbon dioxide gas in air

Decomposers feed on dead plants and animals and release carbon dioxide

Burning coal and other fossil fuels release carbon dioxide into the air

The carbon cycle

Consider the CO_2 produced by your grandparents when they lit their coal stove 50 years ago. Having lain dormant in the coal for perhaps hundreds of millions of years, the carbon atoms were heated up to a high temperature in the stove, causing them to react with oxygen to produce CO_2. Let's follow a single CO_2 molecule to learn more. The CO_2 molecule in the stove escaped out the chimney into the atmosphere, where it was taken on a whirlwind tour of the planet. Sometime during its first decade of travels, the molecule entered the interior of a leaf via photosynthesis, where its two oxygen atoms were stripped away, and its carbon atom became part of the leaf. At the end of the season, the leaf fell to the forest floor. Bacteria or fungi consumed the leaf, reattaching two oxygen atoms to the carbon, and the resulting new CO_2 molecule was released back into the atmosphere. It turns out that in the 50 years since the carbon atom was freed from its lump of coal, it has been part of five different plants. This indicates that the lifetime of a CO_2 molecule in the atmosphere is about a decade. However, this cycle of uptake and release is balanced; it doesn't remove carbon dioxide, it just recycles it. Only processes acting much more slowly (over hundreds or thousands of years), including stirring into the ocean, provide a net removal mechanism.

Release without uptake

While CO_2 has always been released into the atmosphere by natural processes, fossil-fuel burning and deforestation are relatively new sources of atmospheric CO_2. Since this input hasn't been matched by a new removal mechanism, the result has been a continuous rise in atmospheric CO_2 over the last 200 years. Even if we stopped burning fossil fuels today, the return to pre-industrial levels of atmospheric CO_2 would take several centuries.

Rice paddies
Rice paddies are major methane emitters because their flooded soils provide an ideal habitat for bacteria that produce methane as a metabolic byproduct.

Bacterial byproduct

Another culprit in the human greenhouse caper is methane. Methane is a natural gas as well as an anthropogenic one. It is a metabolic byproduct of the microbes that inhabit oxygen-poor environments, such as the black mud of ponds and rice paddies, and the guts of cattle and termites. Because of the low availability of oxygen, a gas they cannot tolerate, these microbes consume organic matter but produce methane (CH_4) rather than carbon dioxide (CO_2) as a byproduct. While some organisms consume methane, most methane released into the atmosphere is removed by chemical reactions that yield CO_2. Thus an increase in methane leads to an increase in carbon dioxide. The average atmospheric lifetime for a methane molecule is about a decade. Agriculture, principally rice cultivation and livestock production, has increased the rate of methane production in recent decades, and atmospheric levels have risen correspondingly.

Agriculture is also the source of nitrous oxide (N_2O), another potent greenhouse gas. N_2O is the natural byproduct of microbes in soils and the ocean, but anthropogenic sources include nitrogen fertilizer, tropical deforestation, and the burning of fossil fuels. These human sources have increased the flow of N_2O into the atmosphere by 40–50% over pre-industrial levels; consequently, the N_2O content of the atmosphere has risen steadily. In the atmosphere, nitrous oxide is slowly broken down by sunlight; the average time an N_2O molecule spends in the atmosphere is a little over a century.

Refrigerants cause global warming

Freons, or chlorofluorocarbons (CFCs), were initially seen as a godsend, because they were efficient, non-toxic refrigerants (i.e., gases used in refrigerators). Only in the late 20th century was it realized that these gases were involved in the destruction of the ozone layer—a region of the stratosphere rich in ozone gas, which protects life on Earth from ultraviolet radiation. Unlike the other greenhouse gases, CFCs, their replacements the HFCs and HCFCs, and other similar gases used as refrigerants and fire extinguishers, have no natural source. Not only are many of these a threat to the ozone layer, they are also strong greenhouse gases.

Global warming potentials

The capacity for the various greenhouse gases to cause climate change differs because of the way in which each molecule interacts with heat. To facilitate comparison, researchers have introduced the concept of "global warming potential," or GWP. A gas's GWP is a calculation of the increase in greenhouse effect caused by the release of a kilogram of the gas, relative to that produced by an equivalent amount of CO_2. GWPs have to be expressed in terms of a time horizon, such as 20, 100, or 500 years, because the different greenhouse gases have quite contrasting atmospheric lifetimes. The table below illustrates the strong greenhouse capabilities of methane, nitrous oxide, and the CFCs. Fortunately, they are being emitted at much slower rates than CO_2, so their overall effect is still less than that of CO_2. Since CO_2 is involved in so many processes, it has multiple lifetimes. However, these processes recycle CO_2 rather than remove it from the atmosphere for good. Its ultimate lifetime, therefore, is considerably longer than that of the other greenhouse gases. As time passes, the relative importance of CO_2 will only increase.

AMOUNT OF GAS IN ATMOSPHERE AS EXPRESSED AS PARTS PER BILLION (ppb)

CO_2 (carbon dioxide)
Amount in atmosphere: 386,000 ppb

CH_4 (methane)
Amount in atmosphere: 1,774 ppb

N_2O (nitrous oxide)
Amount in atmosphere: 319 ppb

CFC-11 (trichlorofluoromethane)
Amount in atmosphere: 0.251 ppb

CFC-12 (dichlorodifluoromethane)
Amount in atmosphere: 0.538 ppb

HCFC-22 (trifluoromethane)
Amount in atmosphere: 0.169 ppb

LIFETIME AND GLOBAL WARMING POTENTIAL OF HUMAN-GENERATED GREENHOUSE GASES

Gas	CO_2	CH_4	N_2O	CFC-11	CFC-12	HCFC-22
Lifetime years	Multiple	12	114	45	100	12
Global warming potential						
20 years	1	72	289	6,730	11,000	5,160
100 years	1	25	298	4,750	10,900	1,810
500 years	1	8	153	1,620	5,200	549

Consider the simultaneous release of a kilogram of carbon dioxide and methane. The atmospheric lifetime of methane (CH_4) is 12 years. In the short term (the first 20 years after the gas is released), methane is a strong greenhouse gas, 72 times more powerful than CO_2. However, because it has a shorter lifetime than CO_2, on century-long timescales it becomes only 25 times as effective, and after 500 years, its potency has been significantly reduced.

Scrapped refrigerators
Although the production of CFC-11 and CFC-12 has been banned by international agreement, these gases still leak out of automobiles and from air conditioners and refrigerators decomposing in landfills.

Isn't carbon dioxide causing the hole in the ozone layer?

This is a common misconception, and confusion about the ozone layer is widespread, having permeated the media and the highest levels of government. The ozone layer is an area of high concentration of ozone molecules in the stratosphere (see p.38). The ozone layer serves an important role: it absorbs most of the solar ultraviolet radiation that bombards Earth.

Without an ozone layer, unhealthy levels of ultraviolet radiation would reach Earth's surface, making the planet largely uninhabitable.

A springtime "ozone hole"—a region where stratospheric ozone concentrations are exceptionally low—has developed in the last three decades over Antarctica. Much of the rest of the world has also experienced a reduction of the ozone layer, albeit to a lesser extent.

Although the short answer to the question of whether carbon dioxide is causing this ozone hole is "definitely not!", the more considered response points to a significant overlap between the factors causing ozone depletion and global warming.

Facts about ozone depletion and its overlap with global warming

- Global warming and ozone depletion are two global problems of paramount importance to society.
- The release and accumulation of human-generated compounds, particularly CFC-11 and CFC-12, are causing the seasonal Antarctic ozone hole and the long-term depletion of stratospheric ozone worldwide.
- CFC-11 and CFC-12 are excellent refrigerants, non-reactive chemically, and non-toxic. This non-reactivity means that they are only very slowly removed from the atmosphere by natural processes. Thus they have lifetimes of many decades in the atmosphere.

Dobson units

100 200 300 400 500

- The reactions that destroy ozone are accelerated when particular clouds called polar stratospheric clouds (PScs) form. These clouds form only under the coldest temperatures of the stratosphere. Although it may seem counterintuitive, the buildup of carbon dioxide in the atmospshere actually results in a cooling of the stratosphere (see p.38). This, in turn, causes more PSCs to form, thereby enhancing ozone destruction. In other words, global warming tends to promote the depletion of the ozone layer.
- During their long lifetimes, CFC gas molecules can leak upward into the stratosphere, where they are fragmented by UV rays. This fragmentation releases highly reactive chlorine atoms that can destroy ozone molecules.
- CFCs are also strong greenhouse gases.
- CFCs are thus double threats to the environment. Their buildup in the atmosphere leads to the destruction of the ozone layer, and at the same time contributes to global warming.

Blue color denotes least amount of atmospheric ozone (ozone hole)

OZONE HOLE OVER ANTARCTICA
The amount of ozone in the atmosphere (measured in Dobson units) dips to very low levels in the Antarctic spring (October), because of the presence of CFC's in the upper atmosphere

Greenhouse gases on the rise

How do we know the composition of ancient air?

Although scientists have only been measuring the amount of greenhouse gas in the atmosphere for the last few decades, nature has been collecting samples for hundreds of thousands of years.

As snow accumulates on the Antarctic and Greenland ice sheets, the pressure of overlying snow compresses the snow into ice. Air trapped in the snow becomes encapsulated in tiny bubbles. Scientists drill into ice sheets, remove samples called ice cores, extract the gas from the bubbles trapped in the ice, and measure the composition of ancient air.

Scientists drill into ice sheets.

Next they remove the ice cores.

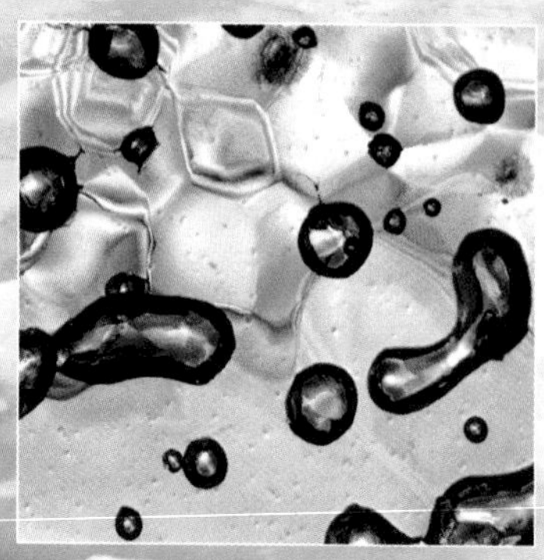

Then they extract the gas from the bubbles trapped in the ice, and measure the composition of these ancient air samples.

The impact of human activity

Together with modern observations, these analyses reveal the unambiguous human effect on atmospheric composition. As the graphs on the right demonstrate, three greenhouse gases—carbon dioxide, methane, and nitrous oxide—have been rising at dramatic rates for the last two centuries. Driven by fossil-fuel burning, deforestation, and agriculture, the recent skyrocketing trends greatly exceed the natural fluctuations of the preceding hundreds of thousands of years. Carbon dioxide and nitrous oxide have risen about 25%; methane has tripled. These gases have a powerful effect on climate, despite the fact that their concentrations are measured in parts per million (ppm) or billion (ppb). You might have to sort through millions of atmospheric molecules to find one of these molecules.

Tragically, the ice-core archive of ancient atmospheres is melting away as climates warm.

CHANGES IN GREENHOUSE GASES: ICE-CORE AND MODERN DATA

Atmospheric concentrations of carbon dioxide, methane, and nitrous oxide are shown here for the last 10,000 years. Concentrations have increased dramatically since the Industrial Revolution.

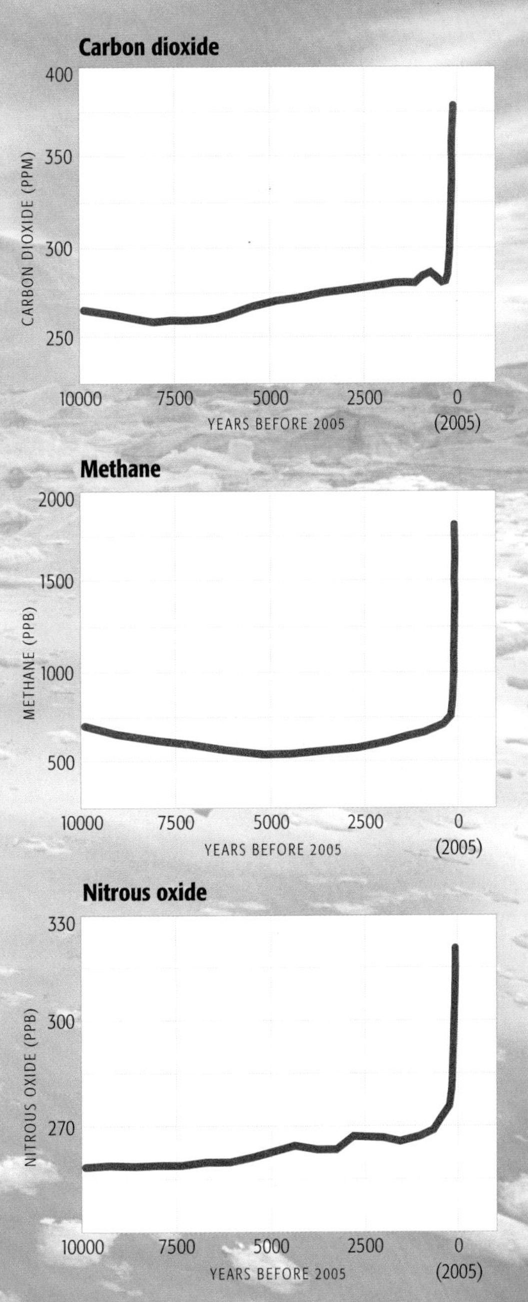

Couldn't the increase in atmospheric CO_2 be the result of natural cycles?

Some talk-radio hosts and other global warming skeptics have claimed that the undeniable rise in atmospheric carbon dioxide (CO_2) levels over the last 50 years could simply be a natural fluctuation.

How do scientists know that it is not?

There are several clues that convince scientists that the CO_2 increase is due almost entirely to fossil-fuel burning.

1 Because fossil-fuel consumption is such an integral part of the global economy, utilization rates are reasonably well known. Looking at these numbers, scientists can determine that fossil-fuel burning can more than account for the recent rise in atmospheric CO_2. In fact, the recent CO_2 rise equates to only half of what has actually been released into the atmosphere. Scientists had to probe deeper to find out what has happened to the other 50%: it has dissolved into the ocean or been taken up by the growth of forests.

No natural source for the CO_2 buildup has been identified.

2 Earth's atmosphere is naturally radioactive, because carbon-14 (radiocarbon) forms in the upper atmosphere. From measurements of tree-ring radioactivity (an indicator of atmospheric radioactivity), we know that this radioactivity remained relatively high prior to the Industrial Revolution. However, it has been decreasing over the last few decades. This indicates that much of the additional carbon driving the rise in atmospheric CO_2 levels is coming from a low-radioactivity or "radiocarbon-dead" source. Volcanoes and the deep ocean are radiocarbon-dead sources, as are fossil fuels. But we know the source couldn't be volcanoes or the deep ocean, because...

3 Carbon atoms exist in three forms, or isotopes. They all have six protons, but each has a different number of neutrons in its nucleus. The most abundant form of carbon, representing nearly 99% of all carbon, is carbon-12, an atom that has six protons and six neutrons. Carbon-12 is stable (non-radioactive); the same carbon-12 atoms that make up your body were once part of an interstellar cloud of material that congealed into our solar system nearly 5 billion years ago. A more rare form of stable carbon is carbon-13, with six protons but

seven neutrons. When plants and algae photosynthesize, they preferentially use molecules of CO_2 that contain carbon-12. Thus when scientists analyze carbon sources that were derived from organic matter, loke the fossil fuels coal and oil, they find that the carbon has a low ratio of carbon-13.

Just as the atmosphere has gradually become less radioactive over time, its ratio of carbon-13 to carbon-12 has been decreasing. This rules out natural, non-plant derived carbon sources, such as volcanoes and the oceans.

Conclusion

The combined trends in the atmosphere's radioactivity and its carbon-13/carbon-12 ratio are satisfactorily explained by only one source: fossil-fuel burning. Although scientists acknowledge that uncertainties exist in our knowledge of global warming, the source of the carbon that has led to the recent buildup of atmospheric carbon dioxide isn't one of them.

Radioactive trees
Tree rings record changes in atmospheric radioactivity over time.

WHAT THE NUMBERS TELL US...

The rise in atmospheric CO_2 since 1800 (graph a) is undeniable. It matches quite closely the increase in human-generated CO_2 emissions, which are quite well known (graph b). The radioactivity of the atmosphere has been decreasing (graph c), implying that the source of the increase is radiocarbon-dead. Also, the ratio of carbon-13 to carbon-12 has been decreasing, implying that the source of the increase was derived from organic matter or plants (graph d). All this points conclusively to fossil fuels as the main cause of the rise in atmospheric CO_2.

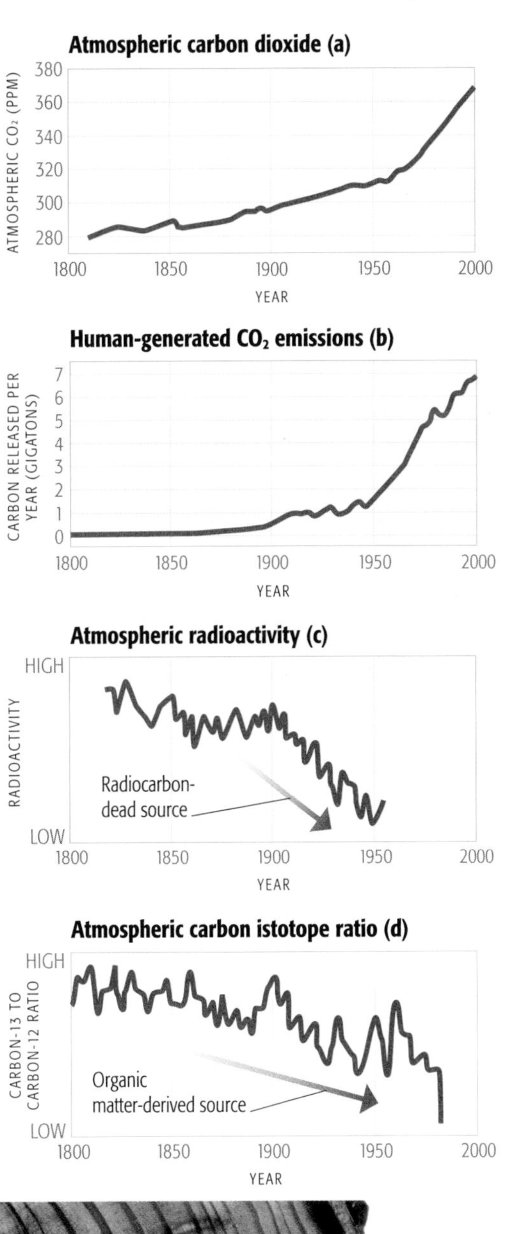

It's getting hotter down here!
Surface temperature observations

Thermometer records have been kept for the past 150 years across much of the globe. During the last few decades, records have been kept almost worldwide. Records include surface air temperatures measured over continents and islands, and sea surface temperatures measured over oceans. By averaging these data across the globe, it is possible to estimate average global temperatures back to the mid-19th century, although uncertainties increase as we look back in time and the data become sparser. We rely on limited instrumental or historical records, supplemented by indirect evidence, to deduce temperature changes prior to the mid-19th century (see p.46).

Accelerated warming

The instrumental temperature record shows that surface warming has taken place across the oceans and land, and that the rate of warming has accelerated over the most recent decades. The average rate of global warming over the full 20th century was slightly less than 0.1°C per decade, but in the past few decades the warming rate has nearly doubled to about 0.2°C per decade. Overall, the average temperature of the globe has warmed from about 13.5°C to 14.5°C since the beginning of the 20th century.

While this warming of roughly 1°C might seem small, it is nearly one-fourth of the estimated change in the temperature of the globe between today and the depths of the last ice age, when New York City was covered by a sheet of ice almost half a kilometer thick. The warming observed so far is only a small fraction of the total

TRENDS IN GLOBAL AVERAGE SURFACE TEMPERATURE

Global temperatures have risen just under 1°C since the mid-19th century, when modern measurements began (see the red line in the graph, which indicates the average rate of change from 1860 to present times). The rate of warming has more than doubled over the past 25 years (see the yellow line).

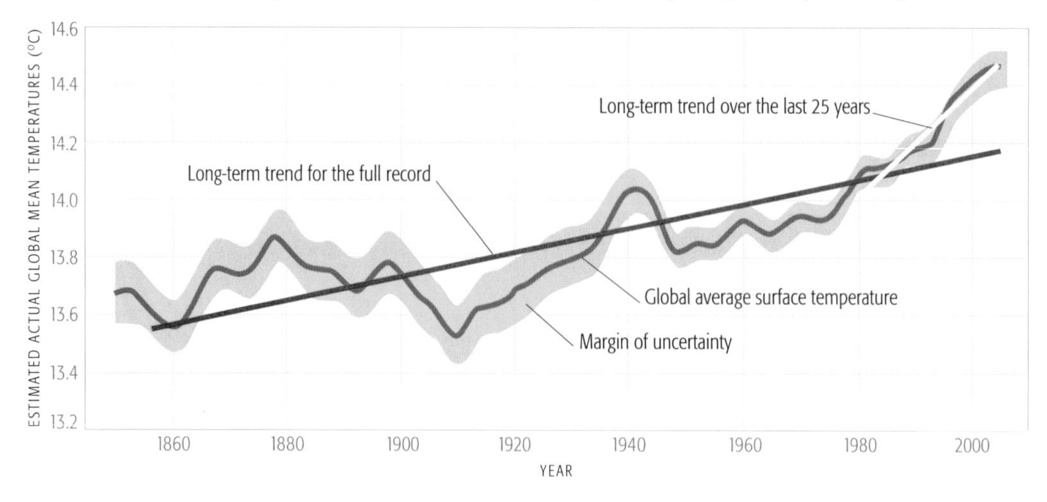

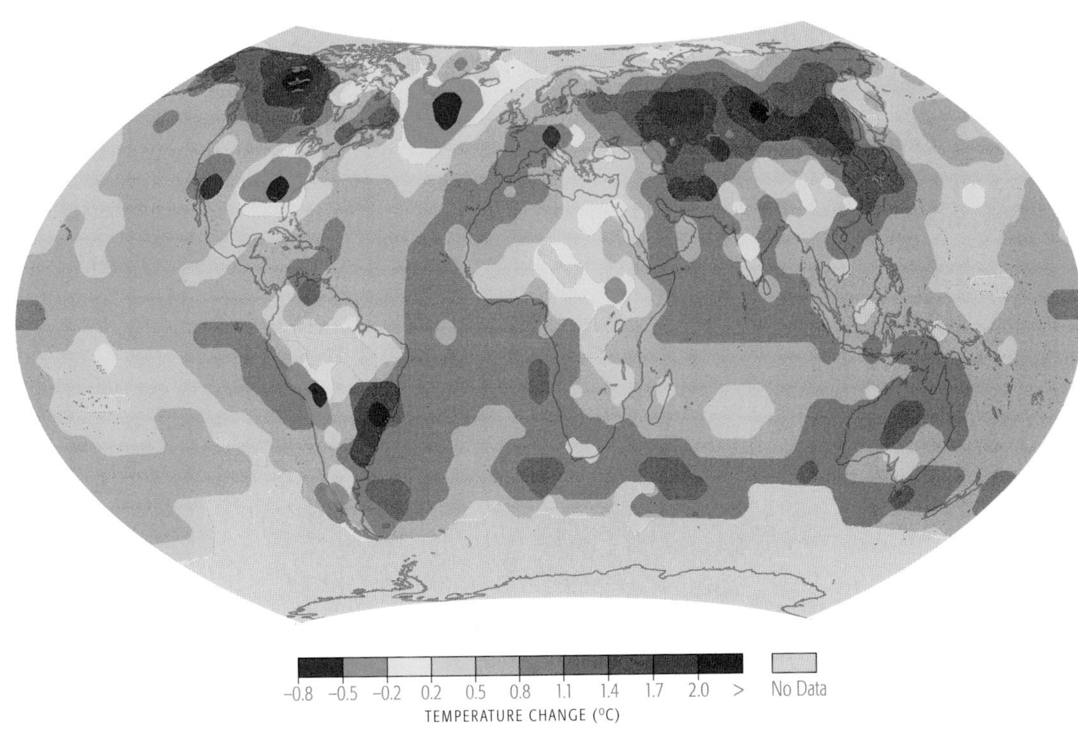

TRENDS IN GLOBAL SURFACE TEMPERATURE 1901–2005

Note that the pattern of warming is not uniform. Land regions, for example, have warmed more than the oceans.

warming expected during the course of the next century, if we continue to burn fossil fuels at current rates (see p.88).

No urban heat bias

Can we trust what the instrumental temperature record is telling us? It is sometimes argued that there may be an "urban heat" bias in the record, due to the fact that cities have warmed up artificially because of their high rate of energy utilization, and the dark, sunlight-absorbing properties of streets and blacktop. Since many records come from urban areas, this bias, it has been argued, may contaminate estimates of global temperature trends.

Scientists, however, have accounted and corrected for these impacts in their assessments of global temperature trends. Furthermore, similar trends are seen when only rural measurements are used.

There are other data, too, with which to counter the skeptics: thermometer measurements indicate that the ocean surface is warming significantly as well. Obviously there is no urbanization impact on sea surface temperatures.

Conclusion

Independent temperature data from the atmosphere (see p.38), the ground, and the ocean sub-surface, combined with evidence such as melting snow, ice, and permafrost (see p.98), rising sea levels, and observed changes in plant and animal behavior make it clear that Earth's surface is warming noticeably.

Is our atmosphere really warming?

At the time of the Third Assessment Report of the IPCC in 2001, there was one body of observational data that appeared to contradict the evidence for global warming. Two measurement sources of atmospheric temperatures over the past few decades—one from microwave measurements made with satellites, the other from weather balloon data—seemed to show that the lower atmosphere (the troposphere) was warming only minimally. This contradicted ground-based thermometer measurements, which indicated substantial surface warming. The inconsistency, argued the skeptics, showed either that the surface data were flawed (see p.36) and the warming trend they indicated was spurious, or that the surface warming was not caused by increased greenhouse gases, since models predicted that the troposphere will warm by as much or even more than the surface if greenhouse gases increase.

In the past few years, however, problems have been found in the older satellite and weather balloon-based assessments. The satellite estimates were compromised by errors that had artificially converted some positive trends into negative trends.

It turns out that the weather balloon data had not been sufficiently quality controlled to eliminate unreliable records. Now that these problems have been identified and dealt with, there is considerably greater agreement between the various atmospheric temperature estimates. The corrected assessments of the satellite and weather balloon data indicate a cooling trend in recent decades in the lower stratosphere, and warming trends at the surface and in the troposphere above it. This is precisely the pattern of atmospheric temperature change predicted by climate model simulations of the response to increased greenhouse gas concentrations.

ATMOSPHERIC LAYERS

Thermosphere gradually thins out until there are no air molecules left

Mesopause is the boundary between the mesosphere and the thermosphere

Stratopause is the boundary between the stratosphere and the mesosphere

Ozone layer (within the stratosphere) absorbs harmful radiation

Tropopause is the boundary between the troposphere and the stratosphere

Thermosphere 87 km and above

Mesosphere 50–87 km

Stratosphere 18–50 km

Troposphere 0–18 km

Sea level

ATMOSPHERIC TEMPERATURE TRENDS

These graphs show observed temperature trends at various altitudes in the atmosphere. (Temperatures represent departures from the 1961–1990 average.)

Lower stratosphere temperature

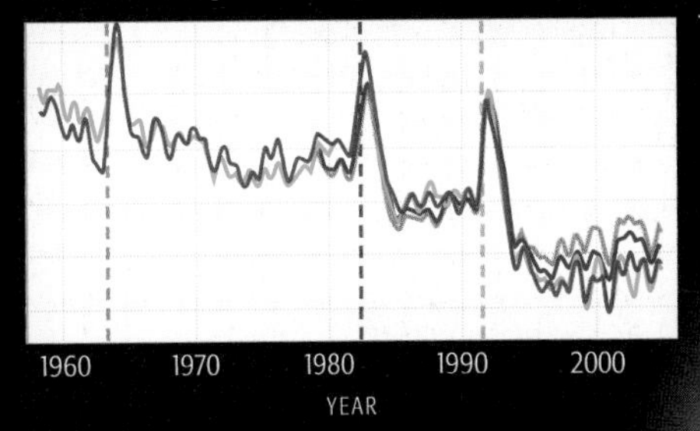

Mid- to upper troposphere temperature

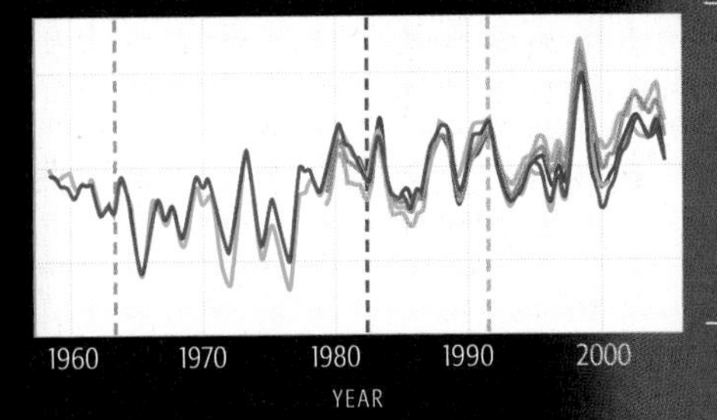

Lower troposphere temperature

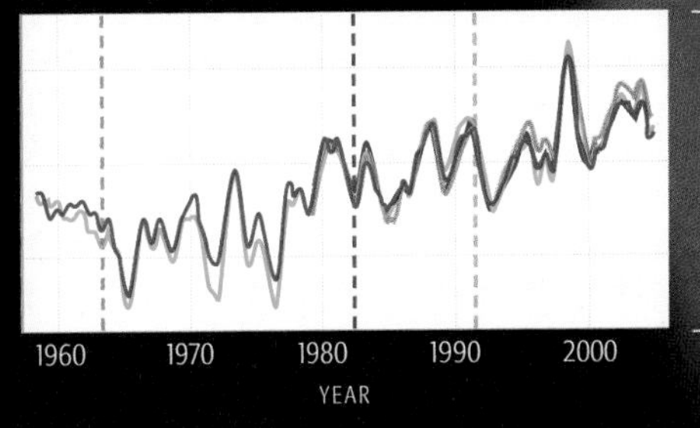

Surface temperature

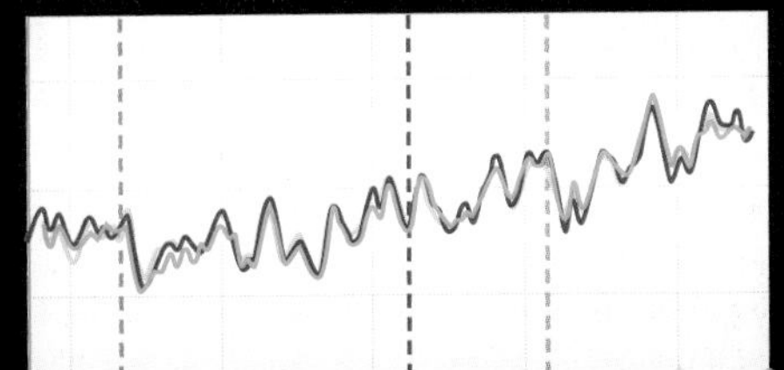

ATMOSPHERIC TEMPERATURE CHANGES

This graphic shows the pattern of late 20th-century/early 21st-century atmospheric temperature changes predicted by climate models. Note that the greatest warming is observed in the tropics and in the lower atmosphere.

KEY

Air temperature analyses from thermometers, satellites, and weather balloons (°C)

Agung volcanic eruption

Back to the future

Deep time holds clues to climate change

When a doctor receives a new patient, a detailed health history is taken. Did the patient suffer these ills previously, and if so, what was the course of the illness? What are the symptoms? What brought on the illness? A physician needs to know these things before making a diagnosis.

Like your body's temperature regulatory system, Earth's climate is self-regulatory, able to resist change but subject to disturbances.

Global warming, in this context, is a planetary "fever"—not particularly high now, but possibly heading toward critical extremes.

If Earth has a planetary fever, then geologists are acting as "geo-physicians," compiling the patient's history by delving into the rock record. Rocks preserve a record of climate history, so studying rocks can tell scientists if there is any link between changing levels of atmospheric CO_2 and climate over Earth's 4.6 billion-year history.

The last two million years

Geologists are pursuing two lines of inquiry: What were the climates of the distant past, and what were the corresponding atmospheric CO_2 levels? By collecting data, geologists have been able to establish a fairly continuous record of ocean and atmospheric temperatures that spans tens of millions of years. As a result, past climate change is quite well understood. Over the last two million years, a time period referred to as the Pleistocene glacial epoch, climates have oscillated between very cold and more livable, like our current climate. By extracting air bubbles from ice cores (see p.32), scientists know that atmospheric CO_2 levels have varied in concert with these temperature swings.

Drill string
The ship's central derrick houses the drill string–thousands of meters of pipe that support the drill bit on the sea floor below.

JOIDES Resolution

The last 65 million years

On longer time scales, we find that climates were generally warmer than today; the last glacial era was over 300 million years ago. The record for the last 65 million years (since the extinction of the dinosaurs) is shown below. Note that the poles were considerably warmer in the past; in the Eocene optimum, for example, alligators and sequoia forests were thriving above the Arctic Circle. The gradual cooling over the past 50 million years is curious—was it caused by declining atmospheric CO_2 levels?

To extend the record of variations in atmospheric CO_2 levels, geologists have applied knowledge from other fields, including biology, biochemistry, and soil science, to develop "proxy" measures (see p.42).

ESTIMATE OF PAST POLAR TEMPERATURE

Sediment cores show that polar temperatures 50 million years ago were up to 12°C warmer than today, and have fallen subsequently in a series of steps.

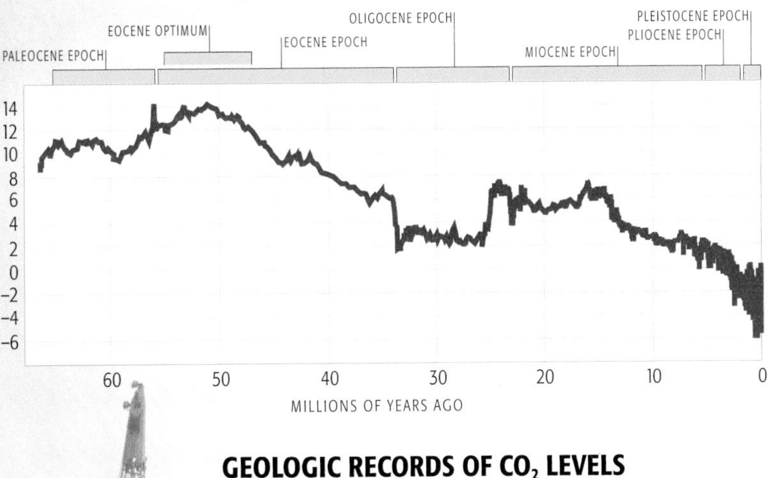

GEOLOGIC RECORDS OF CO_2 LEVELS

Estimates of atmospheric CO_2 levels, based on various proxies, show that levels have fallen over the last 50 million years.

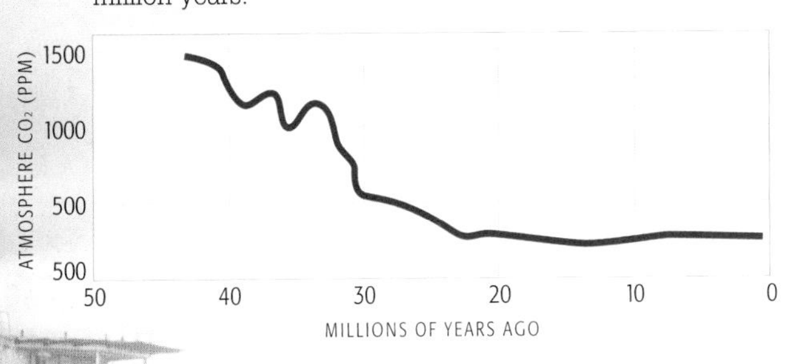

Sediment study
Scientists spend months on the *JOIDES Resolution* recovering sediment cores from the deep-ocean floor.

A fossilized imprint of a prehistoric fern leaf

Living fern leaf

PROXY MEASURE

Proxies are substitutes for what one is really after. Climate scientists study proxies because there are no thermometer records for prehistoric times. Proxies may even include the pores on fossil leaves: when atmospheric carbon dioxide levels are low, plants need more pores on their leaves to bring in more carbon dioxide. Under the microscope, well-preserved fossils reveal the number of pores; comparing this pore density to that of living plants allows scientists to establish carbon dioxide levels in the distant geologic past.

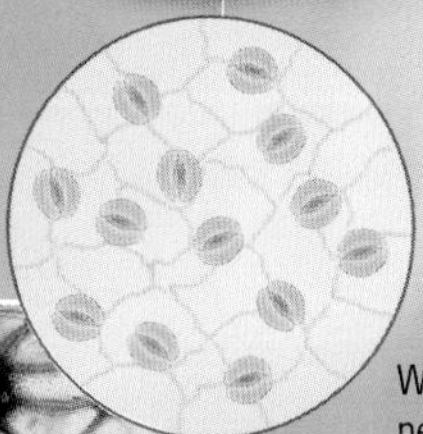

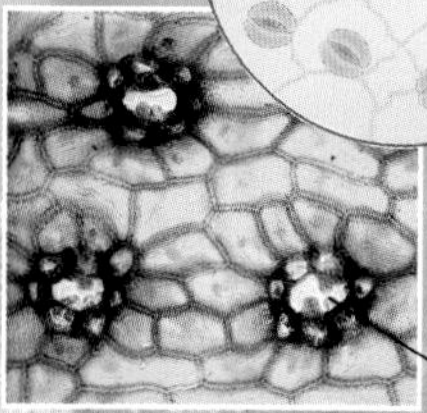

When alive, this tree fern needed a lower density of pores because CO_2 levels were high.

Fossilized leaf pore

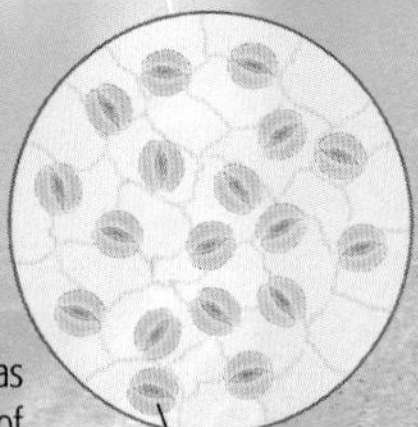

This modern leaf has a higher density of pores because the current atmospheric CO_2 levels are low.

Leaf pore

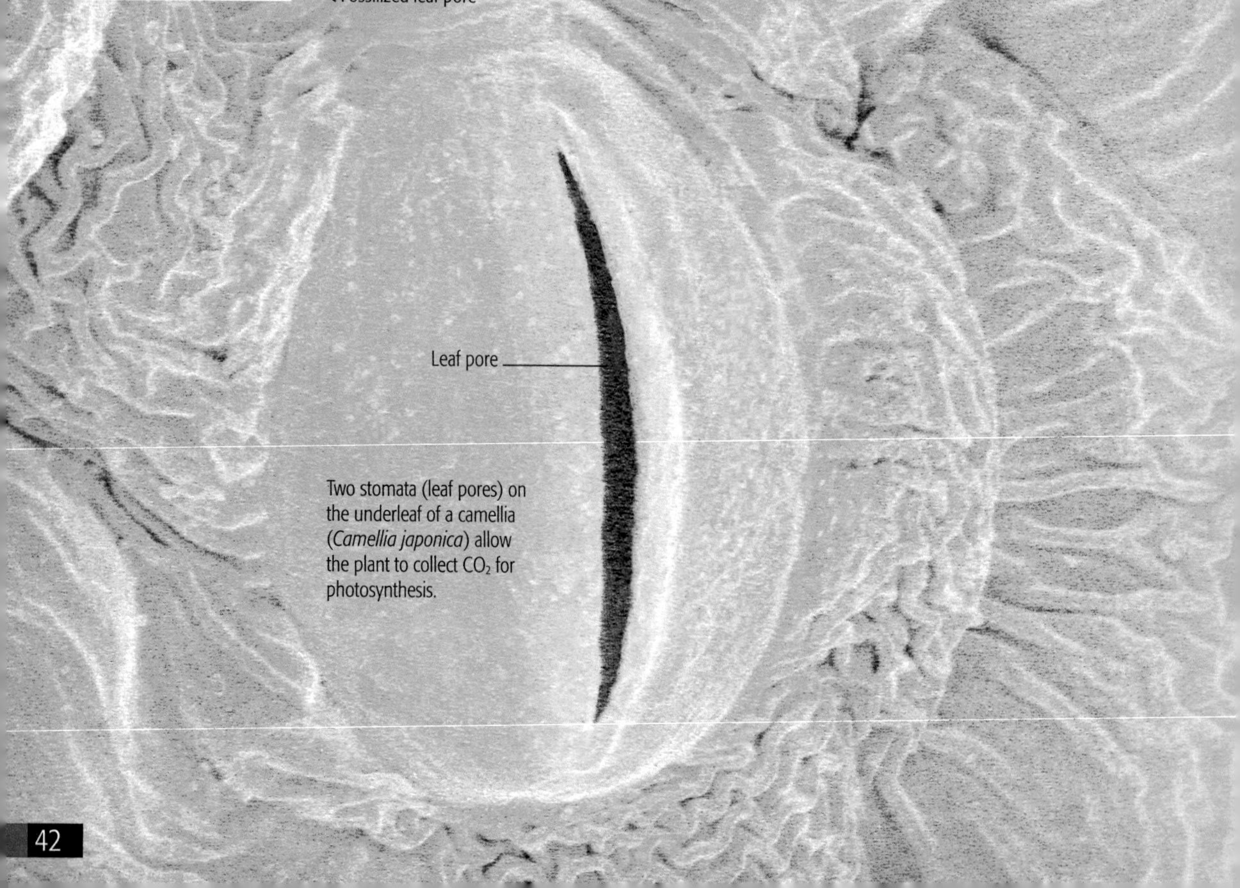

Leaf pore

Two stomata (leaf pores) on the underleaf of a camellia (*Camellia japonica*) allow the plant to collect CO_2 for photosynthesis.

Interpreting the results

Taken together, the geologic records of climate and atmospheric CO_2 levels reveal an expected relationship: when carbon dioxide levels were high, the climate was warm, and vice versa. The correlation isn't perfect, and the mismatches are areas of current research. In particular, the cooling trend from 20 million years ago to the present doesn't seem to be reflected in a reduction in atmospheric CO_2. Are the proxies for climate and CO_2 in error during these times, or is there a third (or fourth or fifth) climate variable that we are neglecting? In this case, the growth of polar ice sheets and their high reflectivity may have provided extra cooling.

Modern measurements of atmospheric CO_2 are 386 ppm. Current estimates of available fossil-fuel reserves translate into the potential for atmospheric CO_2 to rise above 2000 ppm.

If we utilize existing fossil-fuel reserves and do nothing to capture the CO_2 released, the atmospheric CO_2 level will exceed anything experienced on Earth for over 50 million years.

But weren't scientists warning us of an imminent *ice age* only decades ago?

Hollywood License
While human action can have a serious impact on climate, the premise behind the movie *The Day After Tomorrow*—that human activity could send Earth into another ice age—was scientifically flawed.

"Scientists ponder why world's climate is changing; a major cooling widely considered to be inevitable"

THE NEW YORK TIMES
MAY 21, 1975

A myth still perpetuated today in popular critiques of global warming science involves the supposed consensus among climate scientists in the 1970s that Earth was headed into an ice age. As this has obviously turned out not to be the case—the argument typically goes—why should we believe what scientists are saying today about global warming? As is typical with urban myths, there is a small grain of truth to this claim. Ultimately, however, the assertion is incorrect and misleading on several grounds.

It is true that decades ago climate scientists were uncertain about future trends in global average temperature, but there was no consensus that an ice age was imminent, or even that the future trend would be one of cooling. Those ideas were conveyed in sometimes quite alarmist accounts in the popular media during the mid-1970s (e.g., in *Newsweek*, *Time*, and *The New York Times*)—not in scientific publications. Indeed, the National Academy of Sciences concluded in a report published in 1975: "*...we do not have a good quantitative understanding of our climate machine and what determines its course. Without the fundamental understanding, it does not seem possible to predict climate...*"

So why all the uncertainty? First, scientists recognized that Earth was eventually due for another ice age as part of the natural cycles of cold ("glacial") and warm ("interglacial") periods that occur due to slow changes in Earth's orbit around the Sun. Just how far away the next ice age might be was not very well known. Second, scientists were already aware that human impacts on climate included both a regional cooling effect from industrial aerosols and the global warming effect of increased greenhouse gas concentrations due to fossil-fuel burning. But it was still not fully understood which of these effects would dominate in the end. We know now that the cooling from the 1950s to the 1970s was probably due to a substantial increase in the regional aerosol cooling impact, which at the time was overwhelming the greenhouse warming impact in the northern hemisphere (see p.68).

More than 30 years after the ice age scare, much has changed scientifically.

The rate of increase in greenhouse gas concentrations due to fossil-fuel burning has accelerated, while policies such as the Clean Air Acts have dramatically reduced aerosol production in most industrial regions. So the impact of increasing greenhouse gas concentrations has considerably overtaken any aerosol-related cooling. Accordingly, since the 1970s there has been even more warming than during the entire preceding century. Equally important, climate scientists now work with far more reliable models of Earth's climate system than they did 30 years ago (see p.64), and they have a better knowledge of the various natural and human factors that influence climate. It is now clear that a natural ice age is not due for many millennia, and we know that the relative impact of aerosols has been small compared to that of greenhouse gas concentrations in recent decades. We also have considerably more data, and we know that the rate of warming in recent decades is greater than can be explained by any natural factors (see p.68).

NORTHERN HEMISPHERE CONTINENTAL TEMPERATURE TRENDS

When we compare northern hemisphere temperature trends through the current decade with the shorter record that was available in the mid-1970s (inset), we see that the trend is actually toward elevated temperatures, not cooling.

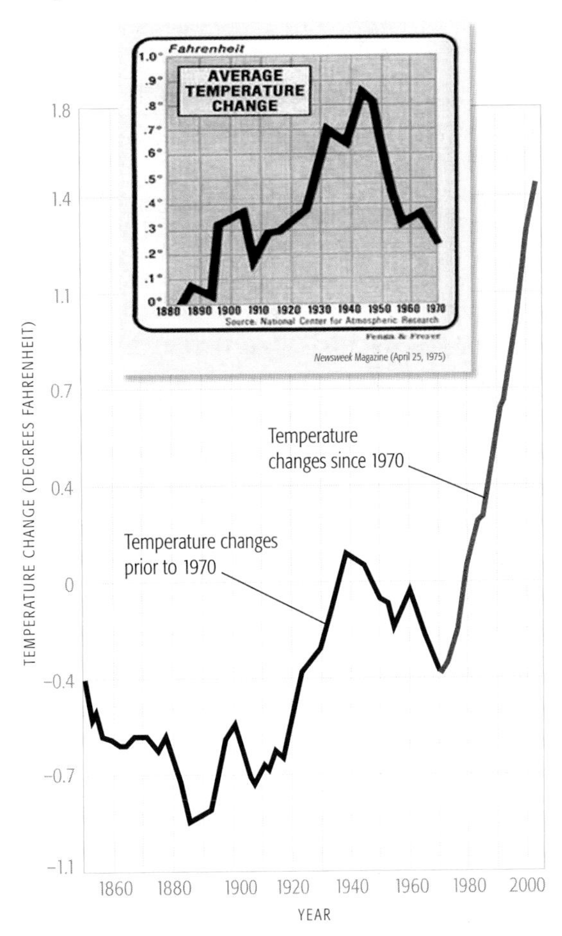

How does modern warming differ from past warming trends?

Some inaccurate accounts of Earth's climate history make reference to a period called the "Medieval Warm Period." It is sometimes asserted, for example, that because Norse explorers were able to establish settlements in southern Greenland in the late 10th century, global temperatures must have been warmer then than now. Supporters of this view also point to the fact that wine grapes were grown in parts of England in medieval times, indicating that local conditions were warmer than they are today. In fact, the ability of the Norse to maintain colonies in Greenland appears to have been related to factors other than local climate (such as the maintenance of vigorous trade with mainland Europe), and wine grapes are grown over a more extensive region of England today than they were during medieval times.

Actual scientific evidence

So how do modern temperatures compare to those in past centuries, based on the actual scientific evidence? Evidence from climate proxy data (see p.42), including tree rings, corals, ice cores, and lake sediments, as well as isolated documentary evidence, indicates that certain regions, such as Europe, experienced a period of relative warmth from the 10th to the 13th centuries, and one of relative cold from the 15th to the 19th centuries (this latter period is often referred to as the "Little Ice Age"). Other regions however, such as the tropical Pacific, appear to have been out of step with these trends.

In fact, the timing of peak warmth and peak cold in past centuries seems to have been highly variable from one region to the next. For this reason, temperature changes in past centuries, when averaged over large regions such as the entire northern hemisphere, appear to have been modest—significantly less than 1°C.

Warming everywhere

Unlike the warming of past centuries, modern warming has been globally synchronous, with temperatures increasing across nearly all regions during the most recent century. When averaged over a large region such as the northern hemisphere (for which there are widespread records), peak warmth during medieval times appears to have reached only mid-20th century levels—levels that have been exceeded by about 0.5 °C in the most recent decades.

Vineyards–in England?
Rows of grapes in the vineyards of Denbies Wine Estate, Surrey, England, UK

Simulations indicate that the peak warmth during medieval times and the peak cold during later centuries were due to natural factors, such as volcanic eruptions and changes in solar output. By contrast, the recent anomalous warming can only be explained by human influences on climate.

NORTHERN HEMISPHERE TEMPERATURE CHANGES OVER THE PAST MILLENNIUM

A number of independent estimates have been made of temperature changes for the northern hemisphere over the past millennium. While there is some variation within the different estimates, which make use of different data and techniques, they all point to the same conclusion: the most recent warming is without precedent for at least the past millennium.

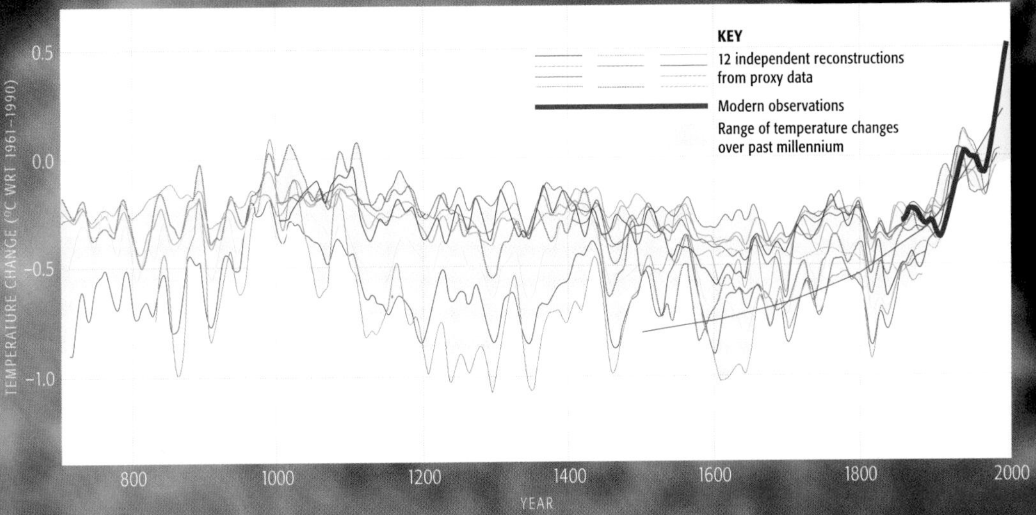

What can a decade of western North American drought tell us about the future?

Much of western North America (the western US, southwestern Canada, and northwestern Mexico) has been gripped by drought since 1999. After a brief respite in late 2004, drought returned in spring 2006, and persisted through summer 2007. Lack of precipitation, reduced river flows, and lower reservoir levels all confirm the seriousness of the drought.

For much of the region, this is the most persistent and severe drought on record.

It is more widespread than the great "Dust Bowl" of the 1930s, which primarily influenced only the central US. What is particularly problematic is that the western US has entered into a pattern of severe drought just as demands for its scarce water resources are skyrocketing, due to increasing irrigation requirements by agribusiness, and dramatically growing populations in Arizona, Nevada, and Utah.

US Drought Monitor categories
D4 Drought – Exceptional
D3 Drought – Extreme
D2 Drought – Severe
D1 Drought – Moderate
D0 Drought – Abnormally Dry

PATTERN OF US DROUGHT IN LATE AUGUST 2003

Extreme drought conditions existed across much of the western US.

WESTERN US PERCENTAGE DRY

The most recent interval of drought is unprecedented in both magnitude and duration.

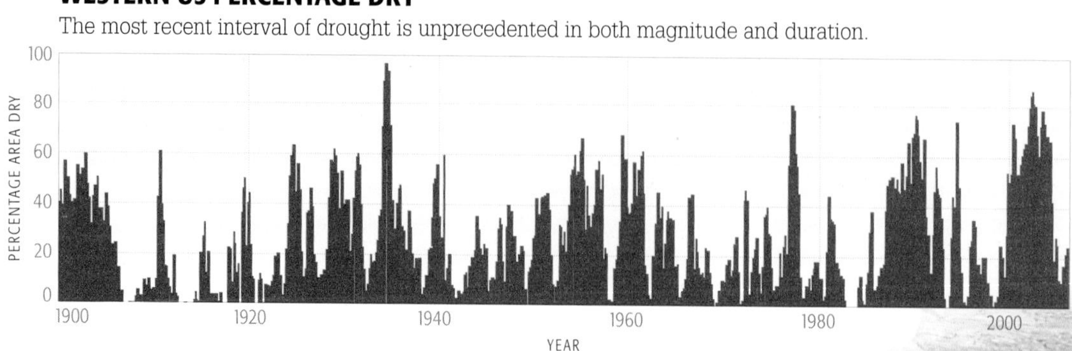

SEA SURFACE TEMPERATURE CHANGES IN THE TROPICAL PACIFIC AND INDIAN OCEANS 1998–2002

These changes are associated with the enhanced North American drought of the past decade. The concentration of warming in the western tropical Pacific and Indian oceans is reminiscent of the east–west temperature contrast typical of La Niña events.

Ocean temperature effects

Is this drought connected with human-caused climate change? That's a tricky question. Recent research ties the persistent drought conditions in western North America to a pattern of ocean surface temperatures in which the eastern and central tropical Pacific are cool relative to the western tropical Pacific and Indian oceans. This is reminiscent of the so-called "La Niña" pattern. Such a pattern has indeed persisted for most of the past decade due, in large part, to anomalously warm sea surface temperatures in the western tropical Pacific and Indian oceans. The brief cessation of drought between 2004 and 2005 was tied to an intermittent "El Niño" event, which represents the reverse pattern. In an El Niño event, the eastern and central parts of the tropical Pacific have warm sea surface temperatures, rather than the western parts. Such a pattern favors wetter conditions over much of western North America (see p.90).

Lake Mead

Lake Mead is a crucial source of fresh water for the more than 20 million people who live in the desert southwest of the US. Projections indicate that a combination of increasing water demand and worsening drought due to climate change threaten to dry up the lake in little more than a decade.

Ocean temperature history

So which of the two patterns will climate change favor more. Warmer conditions in the western tropical Pacific, and drought? Or more frequent and more severe El Niño events, and wet conditions? Paleoclimate records provide a clue: the data indicate that there was persistent drought in the western US during the 11th to 14th centuries, and that relatively wetter conditions prevailed during the 15th to 19th centuries. The later wet conditions appear to have been associated with an El Niño-like pattern, while the earlier dry times were associated with the opposite La Niña pattern. These changes may have, in turn, been driven by the same natural forces—explosive volcanic eruptions and variations in solar output—that were responsible for the relatively globally warm conditions (see p.46) of medieval times (and the relatively cold conditions of the 15th–19th centuries).

If this association of warmer global temperatures with western North American drought continues in response to human-caused global warming, it could portend increasing troubles for the continent in the future.

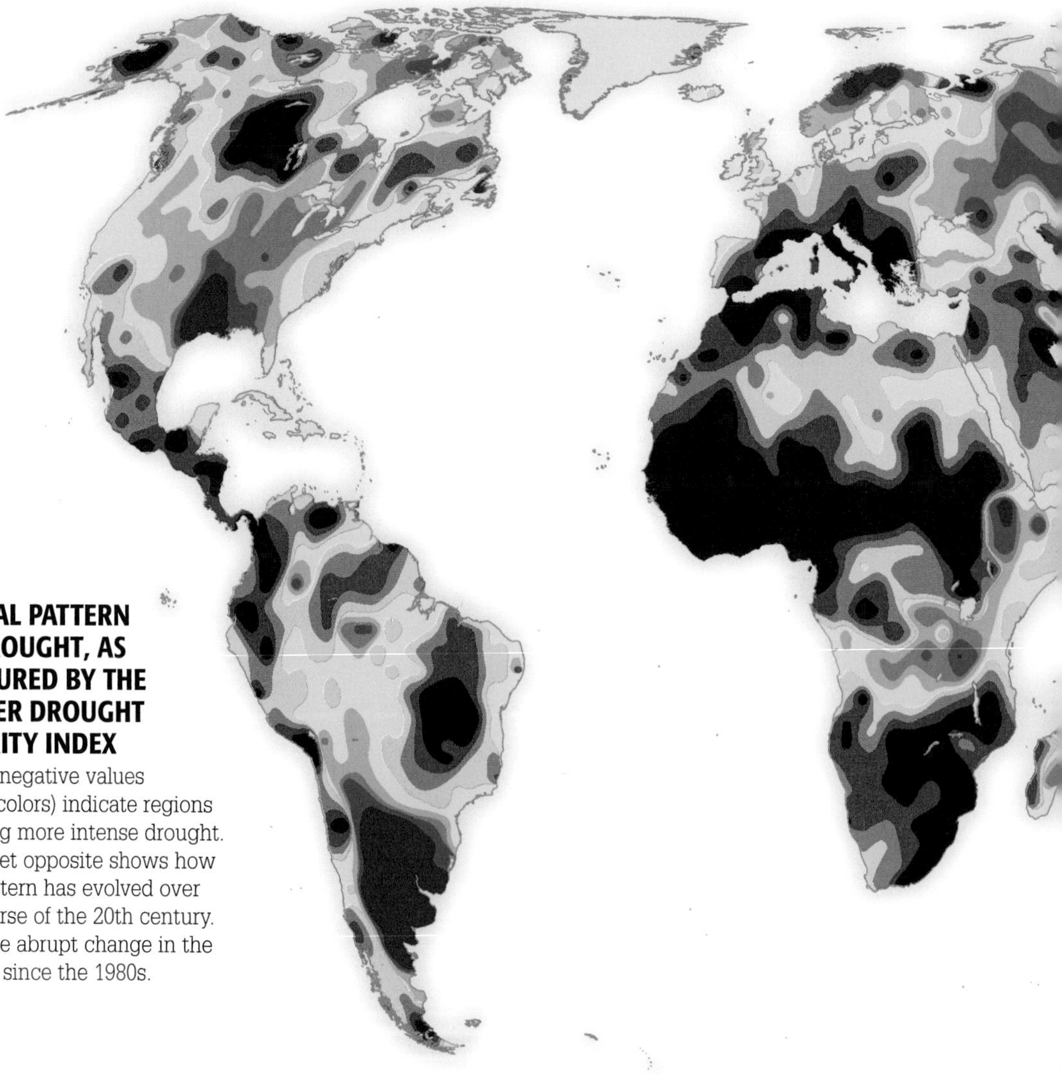

GLOBAL PATTERN OF DROUGHT, AS MEASURED BY THE PALMER DROUGHT SEVERITY INDEX

Higher negative values (warm colors) indicate regions suffering more intense drought. The inset opposite shows how this pattern has evolved over the course of the 20th century. Note the abrupt change in the pattern since the 1980s.

In fact, the drought in the western US is symptomatic of a longer-term, global pattern of increasing drought.

This drought pattern has afflicted much of western North America, the American tropics, most of Africa, Asia, and Indonesia, and parts of Australia. In some cases, the increased drought represents an expansion to higher latitudes of the dry, descending air typically found in the subtropics. In other cases, changes in regional circulation systems such as El Niño (see p.90) and the Asian monsoon appear to be important.

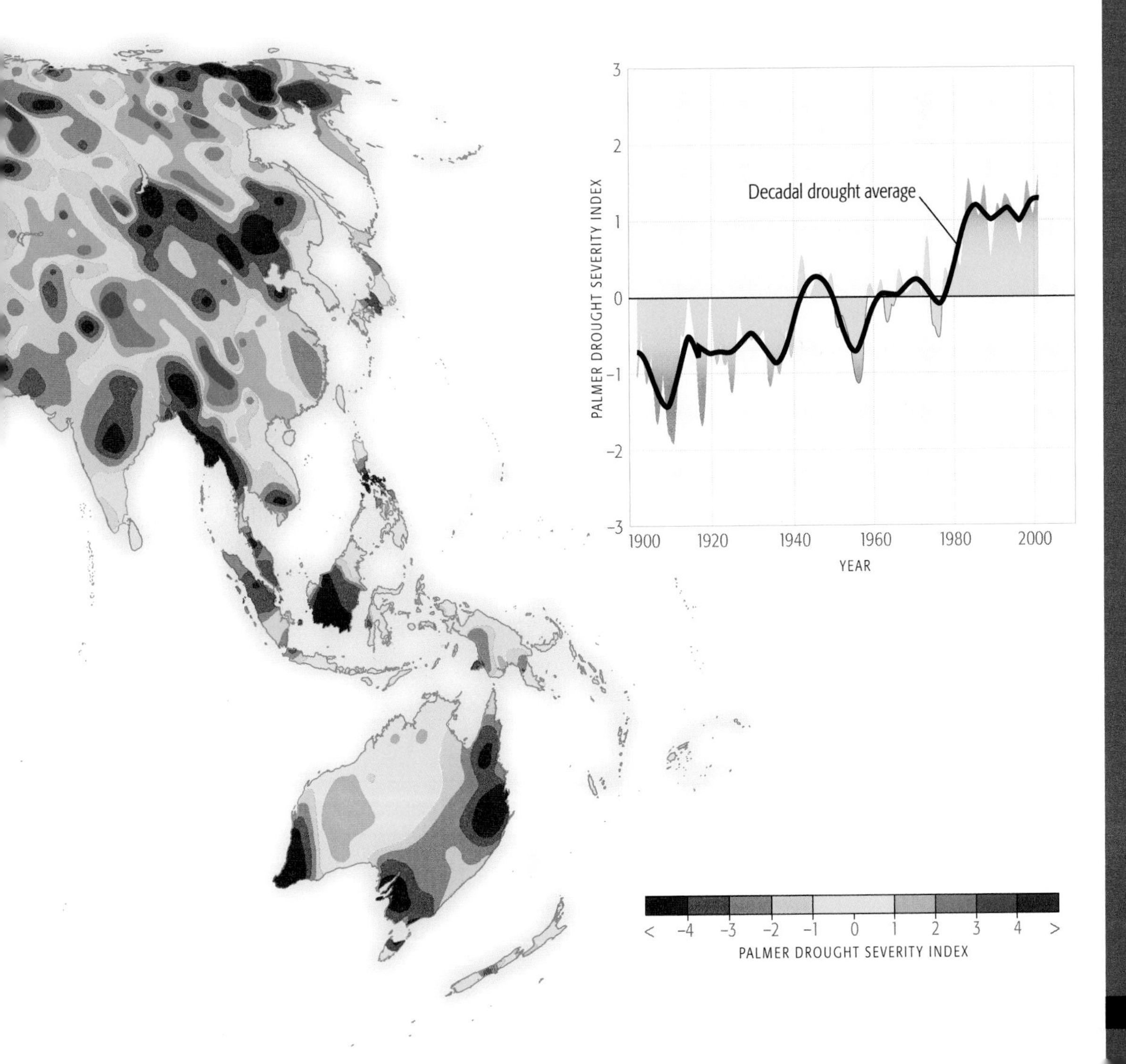

What can the European heat wave of 2003 tell us about the future?

An estimated 35,000 fatalities across Europe were attributed to a unprecedented heat wave during the summer of 2003.

Since the dramatic 2003 heat wave, climate scientists have struggled to determine to what extent this event should be attributed to human-caused climate change. With individual weather events, such as a drenching storm or a destructive hurricane, it is impossible to assess whether climate change played any role. There is simply too much randomness in day-to-day weather to make such an attribution. Heat waves, however, often take place over large areas (e.g., all of Europe or all of eastern North America), and are of extended duration (e.g., several weeks in the case of the 2003 European heat wave). In such instances, it is possible to infer plausible connections to climate change.

In summer 2003, the maximum daily temperature records from June to August were exceeded over an extended area of Europe. In some cases, records several centuries old were broken. These record-setting conditions were part of a coherent, large-scale pattern.

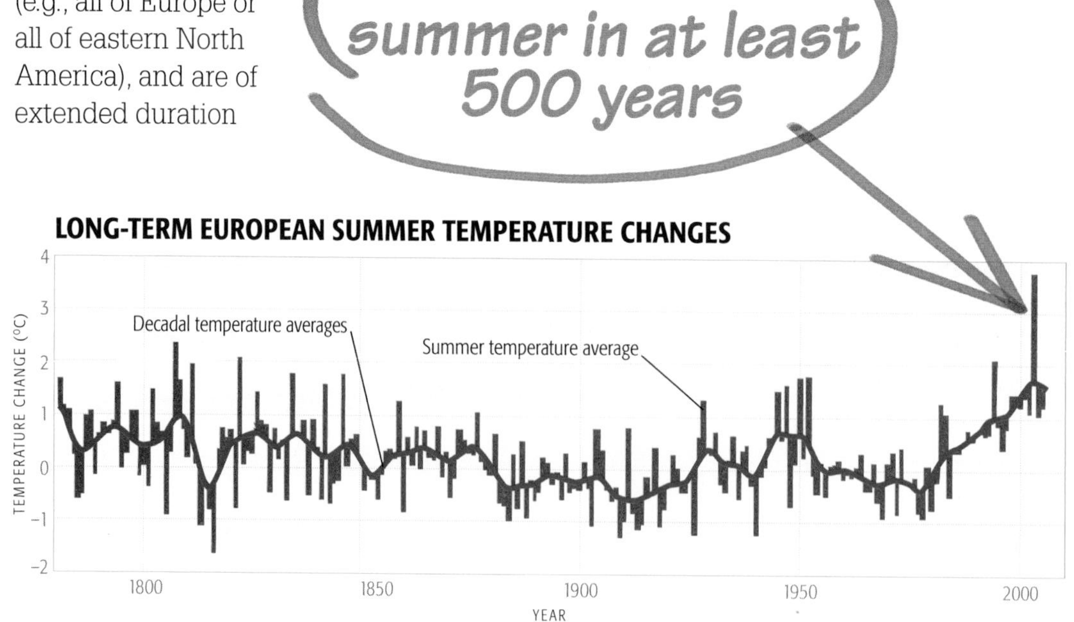

Temperatures for all of central Europe for the entire summer (June–August) of 2003 averaged more than 1.4°C above those for any single year on record.

Not only was the summer of 2003 the warmest since reliable instrumental records began in the late 1700s, but more uncertain longer-term historical records suggest that the summer was the warmest in at least 500 years. This event did not occur in isolation. Instead, it is part of a long-term trend towards more frequent daily extremes of daytime and nighttime warmth over most of the world's major landmasses.

Warm days

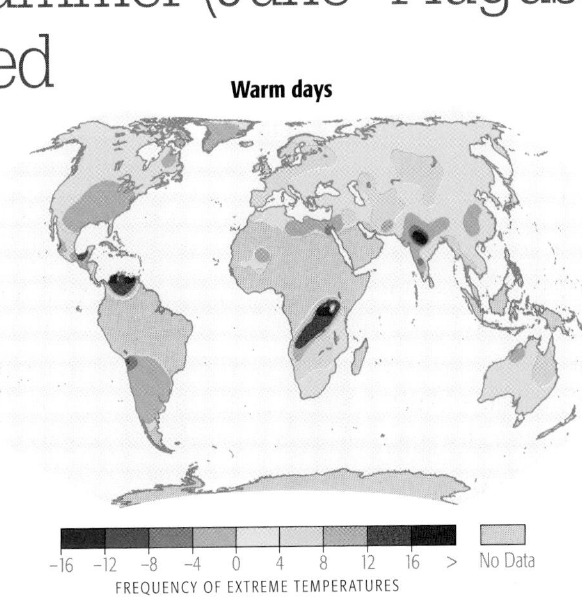

TRENDS IN DAILY EXTREME WARMTH

These spatial maps show the net change from 1951 to 2003 in the number of days and number of nights per decade that qualify as "extremely warm." For these purposes, "extremely warm" is defined as being in the upper 90th percentile of the full (53-year) record. Regions shaded by warm colors represent a positive trend (more extremely warm days/nights), while cold colors represent a negative trend (fewer extremely warm days/nights).

Warm nights

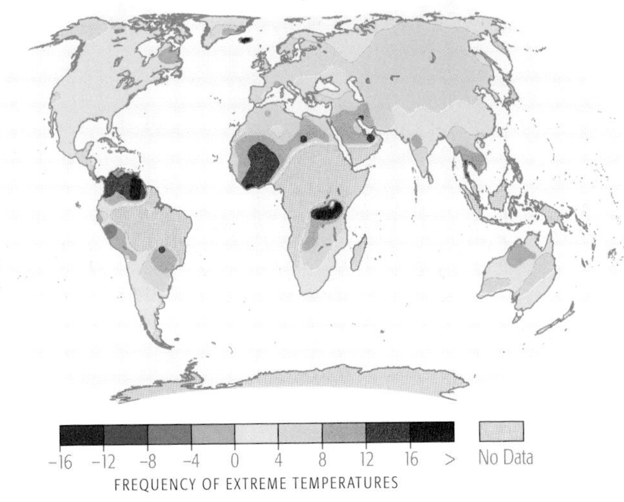

The 2003 European heat wave was associated with an extreme poleward expansion of the high-pressure subtropical zone. This situation is predicted to become more common with human-caused climate change (see p.100).

SUBTROPICAL ZONE EXPANSION

There are two subtropical zones. They lie just to the south of the northern polar jet stream, and to the north of the southern polar jet stream. Climate change is predicted to lead to a poleward migration of the polar jet streams, allowing the dry subtropical zones to penetrate further into mid-latitude regions such as Europe and the US during the summer season.

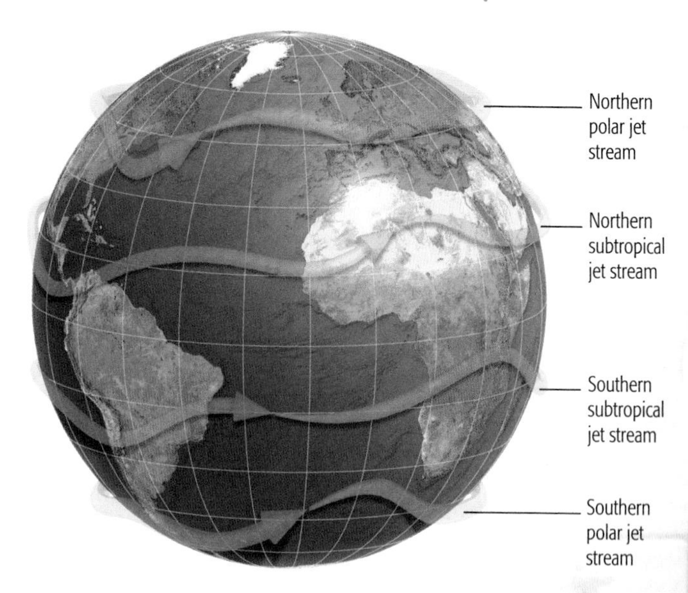

Model simulations indicate that the 2003 event would have been highly unlikely to occur in the absence of human-induced climate change.

In fact, such a likelihood amounts to a roughly once-in-a-thousand-years occurrence. By contrast, when human-induced warming is taken into account, the event is predicted to occur at least once in any given century on average. While this doesn't prove that global warming caused the 2003 European heat wave, it underscores how the increasing occurrence of heat waves is consistent with theoretical expectations, and gives us a taste of what is likely to be in store with future climate change.

Hot in the city
Tourists cool off in London's Trafalgar Square during the 2003 European heat wave.

A tempest in a greenhouse

Have hurricanes become more frequent or intense?

One of the most contentious issues in climate research involves the impacts of climate change on tropical cyclone and hurricane behavior.

A hurricane is defined as an Atlantic tropical cyclone in which winds attain speeds greater than 119 kmh.

Since modern observations of hurricanes from satellites and aircraft reconnaissance missions are only available for the past few decades, it is difficult to determine whether long-term hurricane trends exist, and whether those trends are associated with human-caused climate change.

One measure of tropical cyclone activity is "accumulated cyclone energy," which takes into account the total energy of storms over their lifetime. Such measurements for the last few decades indicate increases in tropical cyclone activity over most, but not all, major formation basins. In some cases, alternative measures of tropical cyclone activity that quantify the destructive potential or "powerfulness" of storms show dramatic recent increases. These increases appear to be closely related to rising sea surface temperatures in the tropical Atlantic (the only region for which long-term records are available).

There is also evidence of a trend toward greater numbers of tropical cyclones in the tropical Atlantic over the past two decades. A global trend is less clear,

The eye of the storm

Hurricane Katrina formed on August 23, 2005. It was the costliest hurricane in U.S. history, wreaking havoc along the Gulf Coast from Lousiana to Alabama, and devastating the city of New Orleans. Katrina was also among the deadliest hurricanes in U.S. history, and was one of three Category 5 Atlantic hurricanes that formed during the record-breaking 2005 season.

however, when the numbers are averaged over all of the major hurricane basins.

Basic theoretical considerations as well as detailed climate-model simulations indicate a likely increase in the average intensity of tropical cyclones and hurricanes in all major formation basins (see p.103). That said, it is uncertain how the actual number of tropical cyclones is likely to be affected by climate change. Accordingly, this remains an area of active climate research.

SEA SURFACE TEMPERATURE VS POWERFULNESS IN TROPICAL ATLANTIC CYCLONES

Sea surface temperatures are closely related to the powerfulness of tropical cyclones. The dramatic increase in cyclone powerfulness over the past decade closely parallels the anomalous rise in sea surface temperature.

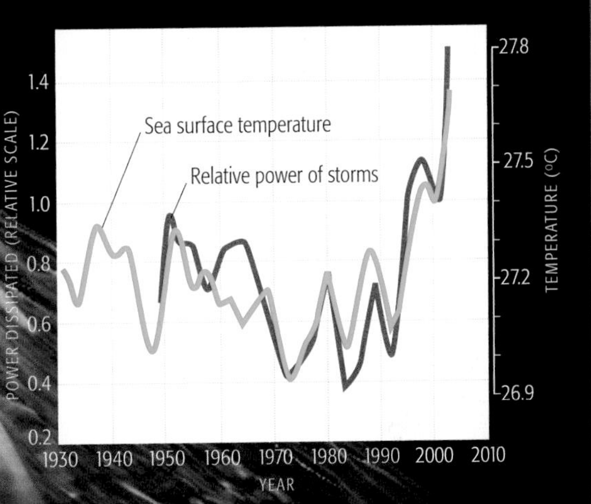

ACCUMULATED CYCLONE ENERGY IN TROPICAL CYCLONE-PRODUCING BASINS

There is a clear upward trend in cyclone energy for the tropical Atlantic in recent years. Trends are less clear for cyclones in the Pacific and Indian oceans. Straight red lines represent the long-term average values of accumulated cyclone energy (ACE) over the 1981–2000 period.

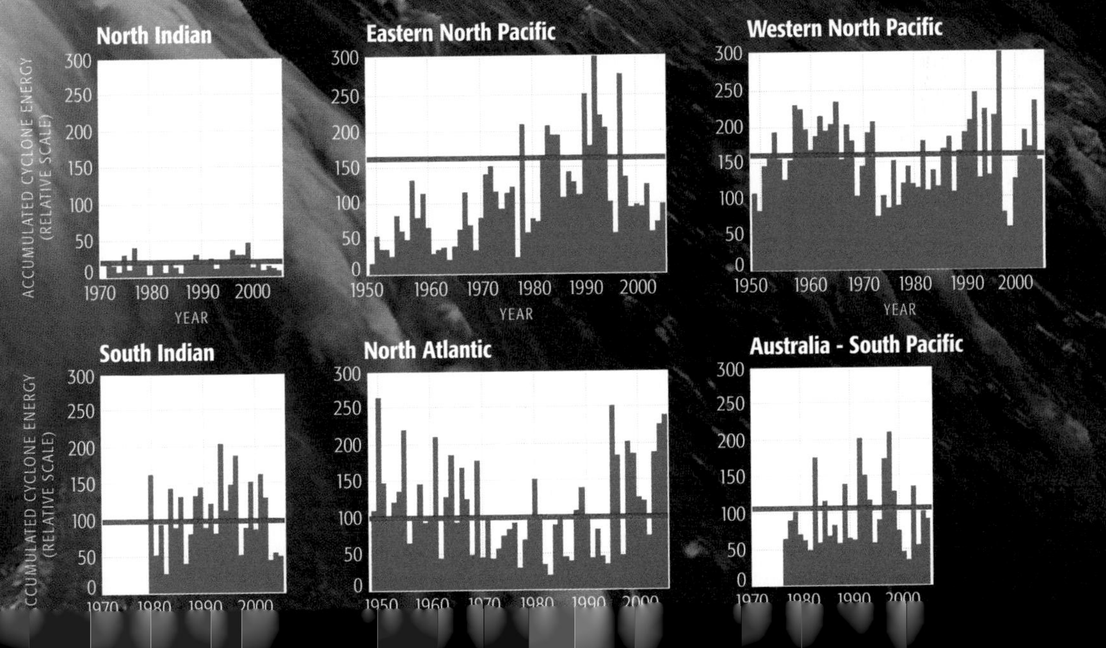

The vanishing snows of Kilimanjaro

An icon of climate change?

There is perhaps no snow-capped peak in the world as iconic as Mount Kilimanjaro, Tanzania, which was immortalized in Ernest Hemingway's *The Snows of Kilimanjaro*. A remarkable group of perennial ice fields can be found at altitudes of roughly 5 km and above atop Mount Kilimanjaro, at a latitude only a few degrees south of the equator.

Like mountain glaciers in many parts of the world, the snows of Kilimanjaro provide a key source of fresh water. The ultimate demise of Kilimanjaro's ice, projected to take place sometime within the next two decades at current rates of decline, consequently poses a threat to the people who inhabit the region.

So are the snows of Kilimanjaro a victim of global warming?

We know that mountain glaciers the world over are disappearing, and that this disappearance is generally related to increased melting due to warmer atmospheric temperatures. In fact, at the altitudes where most tropical mountain glaciers are found, the atmosphere has warmed even more in recent decades than Earth's surface.

1912 (Average area of snow: 12 km^2)

1970 (Average area of snow: 5 km^2)

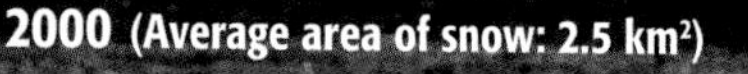

2000 (Average area of snow: 2.5 km^2)

2007 (Average area of snow: 1.5 km^2)

Kilimanjaro has had permanent ice fields for 12,000 years: it is unlikely that their current dramatic retreat is a coincidence.

However, it is not as simple as saying warmer temperatures cause more ice to melt. While some melting has been observed recently on Kilimanjaro, ice loss at these latitudes and elevations occurs mostly through "sublimation" (the evaporation of ice directly into the atmosphere), rather than by melting. The rate of sublimation is influenced by factors in addition to temperature, such as cloud cover and humidity. Also, the amount of ice that exists on mountain glaciers isn't controlled by melting or sublimation alone, but instead represents a balance between the rate of ice loss due to those processes, and the rate of accumulation of ice. Accumulation is determined by the amount of precipitation that falls each year, so it is likely that decreased snowfall in the region has had a significant impact on the extent of the ice fields. The changing precipitation patterns, leading to drier conditions in the region, are tied to larger-scale climate changes. In this sense, human influence on climate is probably responsible for the imminent demise of Kilimanjaro's snows, even if warmer regional temperatures alone are not.

Mount Kilimanjaro
Kilimanjaro, in Tanzania, Africa, is one of a number of locations around the world where ancient mountain glaciers are disappearing before our eyes.

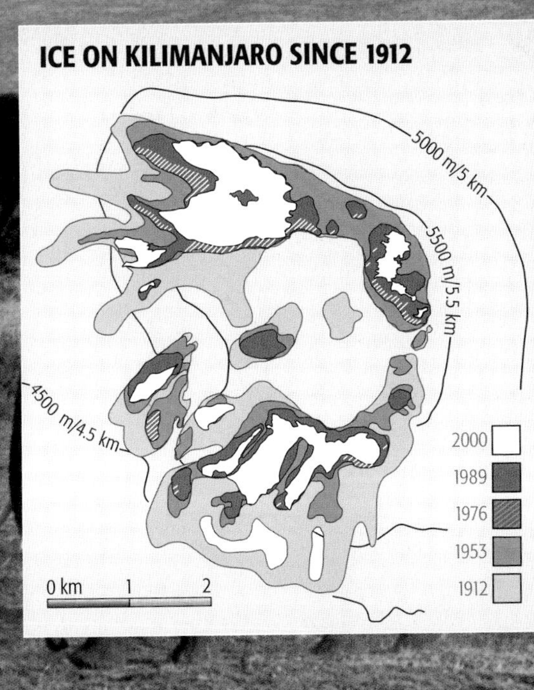

The day after tomorrow

Will the ocean's conveyor belt shut down?

In the 2004 blockbuster movie *The Day After Tomorrow*, global warming causes catastrophic melting of the polar ice sheets. This, in turn, leads to changes in ocean salinity and a shutdown of ocean circulation. The cooling that ensues when ocean waters quit circulating causes super cyclones that wreak havoc worldwide.

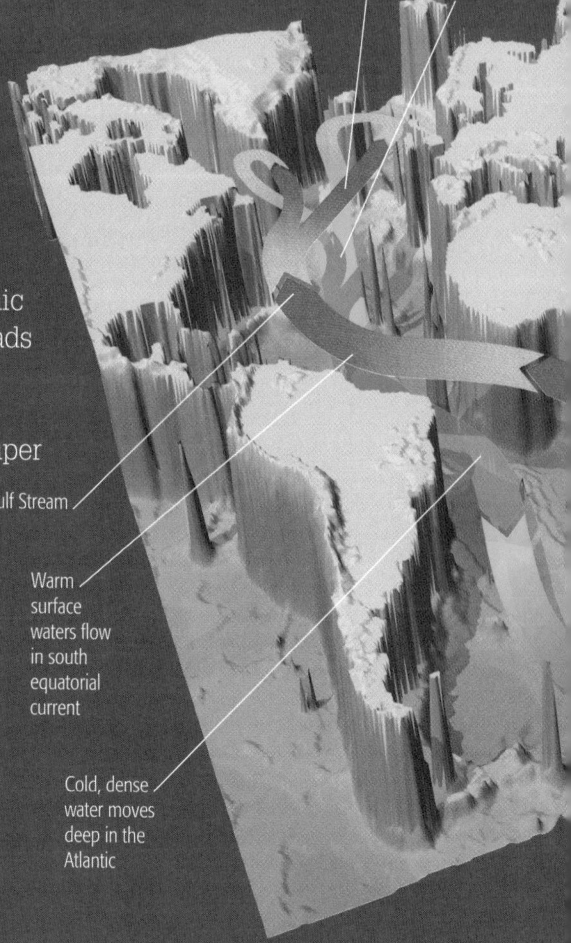

Was the scenario in *The Day After Tomorrow* wholly the brainchild of a screenwriter, or does it have scientific underpinnings?

Although the rate and scale of melting and its consequences are exaggerated in the movie, and in some cases preposterous, the basic notion that the ocean's conveyor-belt circulation is sensitive to the salinity of the North Atlantic is rooted in scientific theory and observation.

If global warming were to increase rainfall in the North Atlantic or cause significant melting of the Greenland ice sheet, the North Atlantic would become less salty. Less-salty water might not sink, which could hamper the flow of the warm ocean current known as the Gulf Stream off the east coast of the United States– and as the North Atlantic Drift along its northeasterly extension (see figure)–and affect global water circulation patterns.

It may have happened before

The climate of northern Europe is milder than that of northern North America at an equivalent latitude. This is the case because air masses move across the Atlantic toward Europe and are warmed by the relatively warm ocean below. A shutdown of the conveyor belt would limit the flow of the warm North Atlantic Drift, cooling northern Europe.

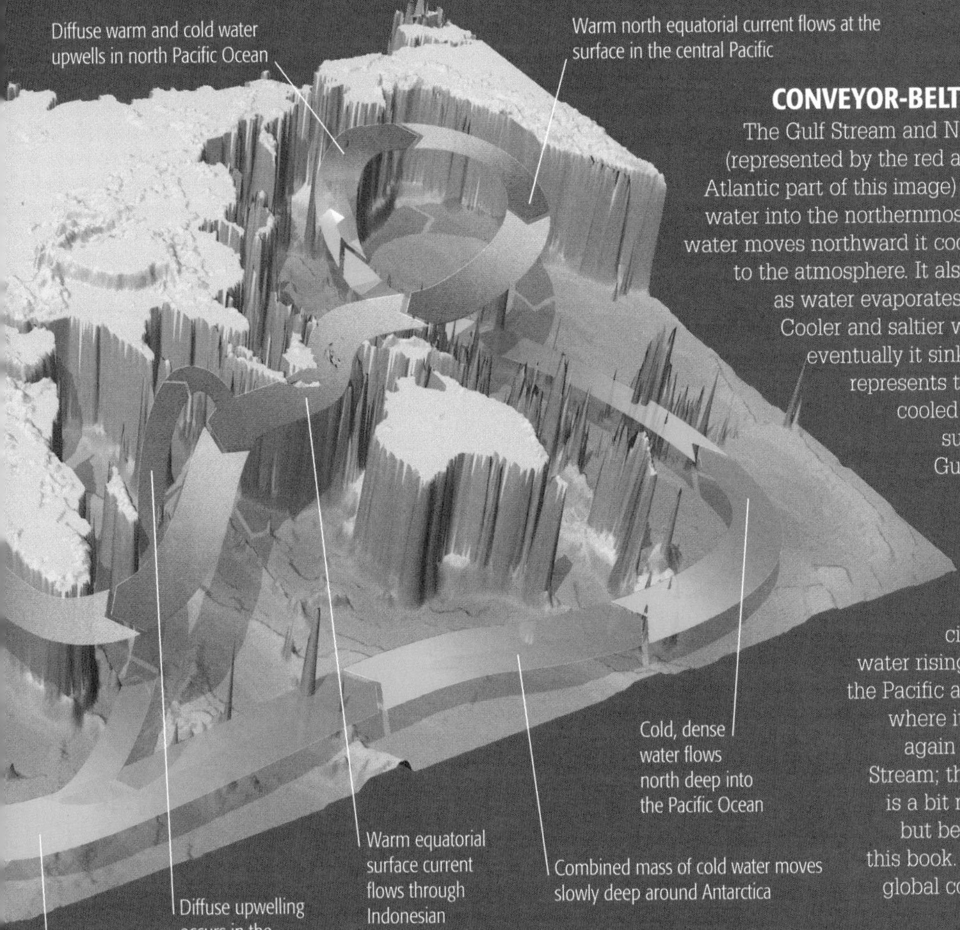

Diffuse warm and cold water upwells in north Pacific Ocean

Warm north equatorial current flows at the surface in the central Pacific

Cold, dense water flows north deep into the Pacific Ocean

Warm equatorial surface current flows through Indonesian archipelago

Combined mass of cold water moves slowly deep around Antarctica

Diffuse upwelling occurs in the Indian Ocean

Atlantic water is joined here by more cold water from near Antarctica

CONVEYOR-BELT CIRCULATION

The Gulf Stream and North Atlantic Drift (represented by the red arrow in the North Atlantic part of this image) carry warm, salty water into the northernmost Atlantic. As this water moves northward it cools, releasing heat to the atmosphere. It also becomes saltier as water evaporates from the surface. Cooler and saltier water is denser, so eventually it sinks (the blue arrow represents the water that has cooled and sunk). At the surface, more warm Gulf Stream water is drawn northward to replace it. The pattern of the ocean's conveyor-belt circulation involves water rising to the surface in the Pacific and Indian oceans, where it warms and once again becomes the Gulf Stream; the actual situation is a bit more complicated, but beyond the scope of this book. One circuit of the global conveyor belt takes 500–1000 years.

Paleoclimatologists have documented an abrupt climate change that occured the end of the last ice age, 12,000 years ago. They think it can be attributed to a temporary shutdown of the North Atlantic ocean circulation. This event was characterized by an abrupt return to glacial climate conditions in northern Europe. The likely cause is a massive infusion of glacial meltwater, which caused a shutdown of the ocean's conveyor-belt circulation.

Climate models predict that under most climate change scenarios, the Greenland ice sheet may be devastated in the coming centuries. The rate of its melting is most uncertain, and it is dependent on the nature of poorly known feedback loops within the ice-sheet system that could accelerate the melting dramatically (see p.98). Even so, the latest climate models indicate that the rate of injection of fresh water into the North Atlantic will not be sufficient to cause a complete shutdown of the conveyor-belt circulation. Some slowdown will likely occur, though, and the resulting cooling of the North Atlantic and northern Europe might offset some of the effects of global warming. The real "Day After Tomorrow," albeit bleak, will bear little resemblance to the Hollywood version.

The last interglacial
A glimpse of the future?

Driving south on highway U.S. 1 from Miami, Florida, you pass a road sign for the Windley Key Fossil Reef Geological State Park, the site of a former limestone quarry. Here, you are at the highest point in the Florida Keys—islands that 125,000 years ago were an impressive chain of coral reefs. At that time, sea level was 6 m above where it is today, most likely because the Greenland ice sheet was smaller and much of the water it comprises today was at that time in the ocean, not locked up in glacial ice. The climate then began to cool, and the reefs became exposed as the sea level fell when water from the ocean evaporated and froze to form ice sheets in the northern hemisphere. At the height of the last glaciation, the seas had retreated 10 km, and the Keys stood 120 m above the ancient sea level. Now they are barely a few meters above sea level, with Windley Key at the highest elevation (6 m). These fluctuations in sea level are the result of the 40-thousand- to 100-thousand-year "glacial–interglacial" cycles of the last 2 million years. These cycles occur in conjunction with the repeated growth and demise of the almost 2-km-thick North American ice sheets that covered most of Canada and the northern United States, as well as other smaller ice sheets in northern Europe.

Will the Florida Keys turn back to coral reefs once again?

Climate in the last interglacial
Indeed, with modern global warming causing sea levels to rise again, low-lying areas like the Florida Keys are gradually disappearing (see p.98). To understand the future implications of these changes, scientists naturally turn to time periods with analogous conditions—like the last interglacial—for clues to questions such as how much the Greenland ice sheet is likely to shrink.

Florida Keys
Highway U.S. 1 extends over a fossil coral reef that formed 125,000 years ago.

A wealth of information on the climate of the last interglacial has been collected in recent decades. Ice cores (see p.82) reveal that atmospheric CO_2, methane, and nitrous oxide were then close to pre-industrial levels. Coastal ocean temperatures were generally warmer than today, there was considerably less sea ice surrounding places like Alaska, and boreal forests had overtaken regions that are now tundra in Siberia and Alaska. Arctic summers were warmer than pre-industrial summers by 5°C.

Earth's orbit affects temperature

Why was the last interglacial so warm—warm enough to melt the Greenland ice sheet? We know from ice-core data that it wasn't due to higher CO_2 levels. The answer is that the northern hemisphere was receiving 10% more solar radiation than it does today, not because the Sun was brighter, but because Earth's orbit around the Sun was different than it is today. Earth's orbit changes slowly and regularly in response to the tug of the Sun, Moon, and the large planets, especially Jupiter. The orbital configuration 125,000 years ago provided more direct rays to northern latitudes in the summer.

Even though this wasn't a greenhouse warming event, and thus not a direct analogy, the information we can learn from the last interglacial suggests that the Greenland ice sheet is subject to considerable shrinkage from relatively subtle changes.

EARTH'S CHANGING ORBIT

Earth's orbit and rotation are not constant, but change cyclically over many millennia. Its orbit varies from elliptical to circular, and the planet also tilts about its axis and wobbles as it rotates. Over time, these changes affect temperature, and they can be correlated to climate swings and glacial–interglacial cycles.

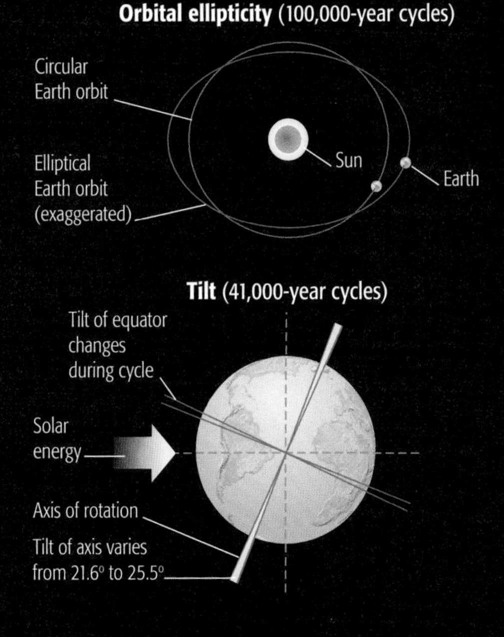

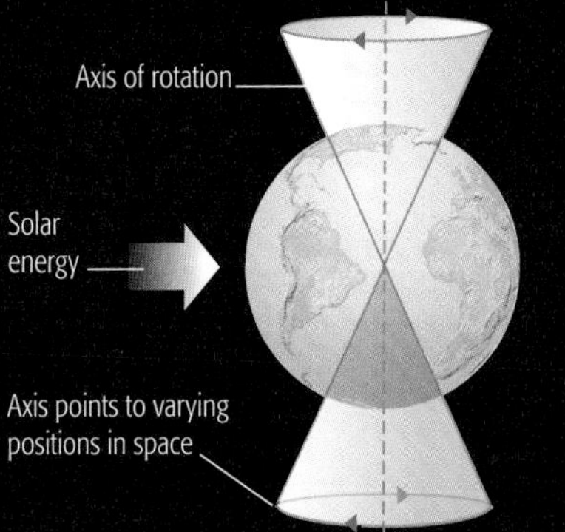

How to build a climate model

Building a model of Earth's climate is a very challenging endeavor; the climate is governed by many complex physical, chemical, and biological processes and their interactions. Earth's climate can be thought of as a system in which fluids with very different properties (Earth's oceans, atmosphere, and ice "cryosphere") interact with each other, as governed by the laws of fluid dynamics, thermodynamics, and radiation balance. There are many further complications, however, that must be taken into account in modeling Earth's climate. For example, life on Earth (the "biosphere"), which includes both plants and animals, plays a key role. The biosphere is involved in the global recirculation of water and carbon, and it influences the composition of the atmosphere and Earth's surface properties—all of which impact on climate.

Simple climate models

The simplest models ignore the three-dimensional structure of Earth, atmosphere, and oceans, and simply focus on the balance between incoming solar energy and outgoing terrestrial ("heat") energy. It is the balance between these incoming and outgoing sources of energy that determines temperatures on Earth. Even in these simple models, the greenhouse effect (see p.22) must be accounted for. This is usually accomplished through a modification that represents the way heat is absorbed and emitted by the atmosphere. It is also essential, even in simple models, to account for "feedback loops" that can either amplify (positive feedback) or diminish (negative feedback) the impacts of any changes (see p.24). In most climate models, the net impact of feedbacks roughly doubles the magnitude of the expected warming or cooling response to imposed changes.

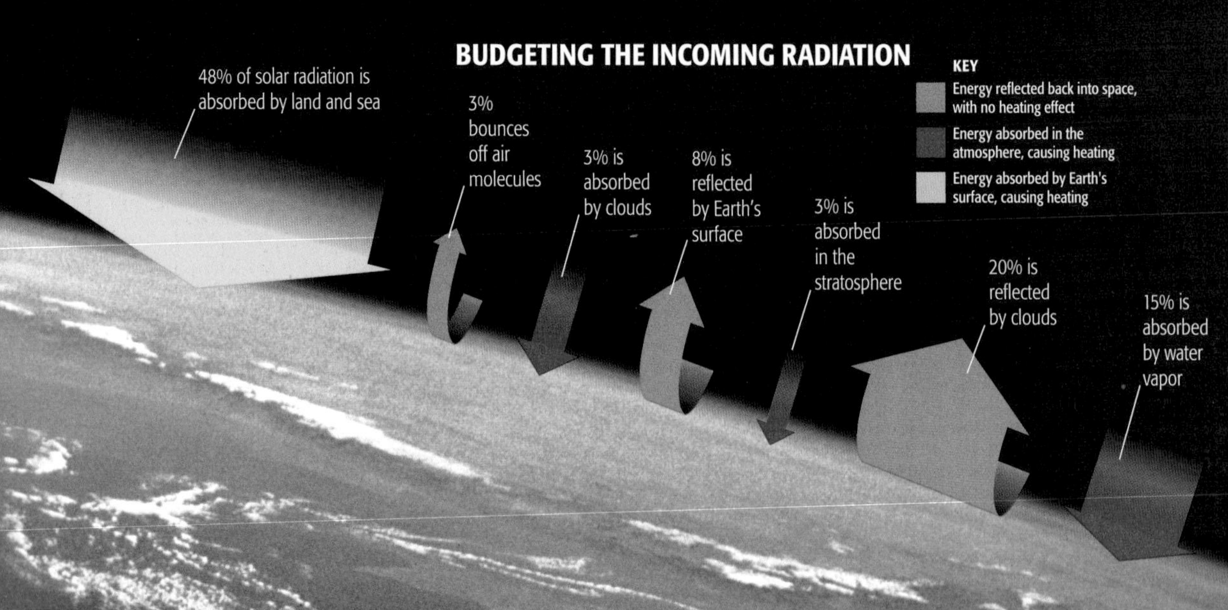

The most complex climate models, referred to as "General Circulation Models," take into account the full three-dimensional structure of the atmosphere and oceans, the arrangement of the continents, the details of coastlines and ocean basins, and the surface topography. These models calculate not only surface temperatures, but also other important climate variables, such as precipitation, atmospheric pressure, surface and upper level winds, ocean currents, temperatures, and salinity. All this is accomplished by breaking the oceans and atmosphere into many small grid boxes, and using the underlying physical, chemical, and biological relationships to calculate values for the properties of each box and the interactions between the different boxes.

Should climate model predictions be trusted?

Current climate models do a remarkably good job of reproducing key features of the actual climate such as the jet streams in the atmosphere, the seasonal band of rainfall and cloudiness that migrates north and south of the equator, and even the complex internal climate oscillation associated with the El Niño phenomenon (see p.90). They also closely reproduce past changes (see p.68). We therefore have good reason to take their predictions of possible future changes in climate seriously.

Atmosphere is divided into 3-D grid boxes, each with its own local climate

Air in grid boxes interacts horizontally and vertically with other boxes

Influence of vegetation and terrain is included

Water in oceanic grid boxes interacts horizontally and vertically with other boxes

Oceanic grid boxes model currents, temperature, and salinity

COMPLEX CLIMATE MODELLING

A climate model is a computerized representation of the Earth including its atmosphere, oceans, and various other components, based on a three-dimensional global grid. The model then applies these changes to its virtual world, to see what effect they have on the climate.

LOST ENERGY

Only 48% of incoming solar energy reaches Earth's surface to heat the continents and oceans. Nearly one-third of the total energy that encounters the atmosphere is immediately returned to space—reflected by clouds or air molecules. Additional radiation is reflected by Earth's surface, especially in icy regions. The remainder is absorbed by stratospheric gases and tropospheric clouds and water vapor.

Profile: James Hansen

James Hansen is a well-known climatologist and Director of NASA's Goddard Institute for Space Studies (GISS), a major climate-modeling center. Hansen, a member of the prestigious U.S. National Academy of Sciences, came to prominence during his congressional testimony in the hot summer of 1988. On June 23 of that year, in front of the US Congress, Hansen became the first scientist to publicly testify that society was already beginning to witness the effects of human-caused climate change. His testimony appears prescient when we look at what has happened since his now-famous pronouncement.

Projections proved correct

During that testimony, Hansen presented projections about likely future warming in terms of three possible scenarios (see chart at right).

The projections Hansen made in 1988 have proven to be a key validation of the models used by climate scientists.

Comparing Hansen's predictions of future global temperature changes with actual temperature observations reveals a remarkable match. This shows that successful projections of global temperature can be made decades in advance.

In 2001, Hansen received a Heinz Award in the Environment for his "exemplary leadership in the critical and often-contentious debate over the threat of global climate change." He remains a productive climate scientist, and is actively involved in efforts to educate the public and policy makers about climate change and the threat it poses to society.

HANSEN'S THREE PROJECTED GLOBAL WARMING SCENARIOS

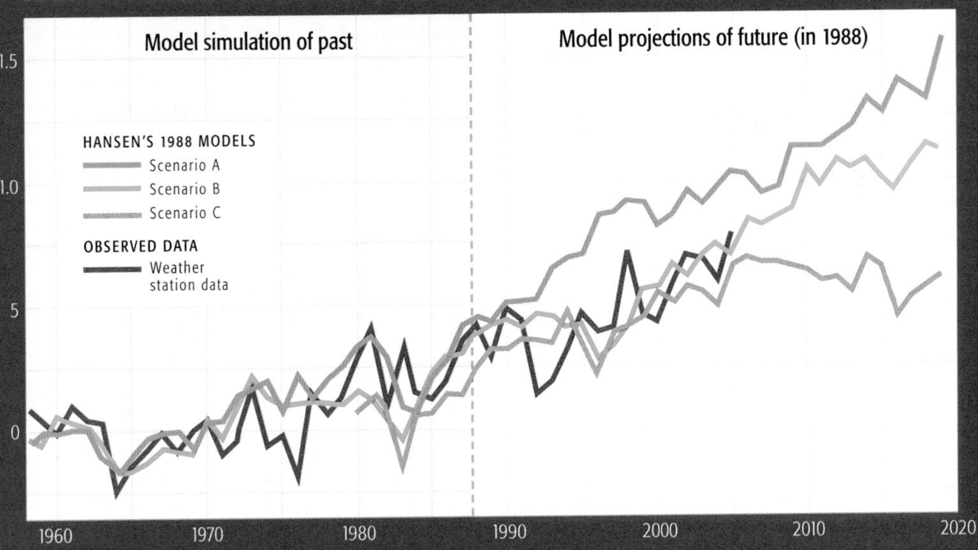

This graph shows climate model simulations performed by James Hansen in 1988, updated to indicate how projected global average surface temperatures have compared with the actual observed temperatures. Scenario B (the blue line) corresponds roughly to the actual greenhouse gas emissions pattern that society has followed since 1988.

Comparing climate model predictions with observations

The average annual temperature of the planet is not expected to be constant in the absence of human influence. Variations in solar energy input warm and cool the planet on yearly, decadal, and longer timescales, while volcanic eruptions cool the planet from months up to years after a major eruption.

Using climate models

We can use climate models to calculate how natural variations in solar energy and volcanic aerosols alone would have driven climate change over the last century if there hadn't been any human influence (graph 1), and compare the result to the observed record (graph 2). Then we can include the human influence (graph 3), and see if the model predictions fit the observed data better (graph 4). The temperature deviations in the graphs below are expressed relative to the average temperatures from 1901 to 1997. The good fit between actual observations and models that take into account human actions and actual data gives us confidence that we can predict future climate responses to fossil-fuel burning.

PREDICTED/OBSERVED CLIMATE TRENDS

1 Predicted temperature trends from models, taking into account the impacts of natural forces alone

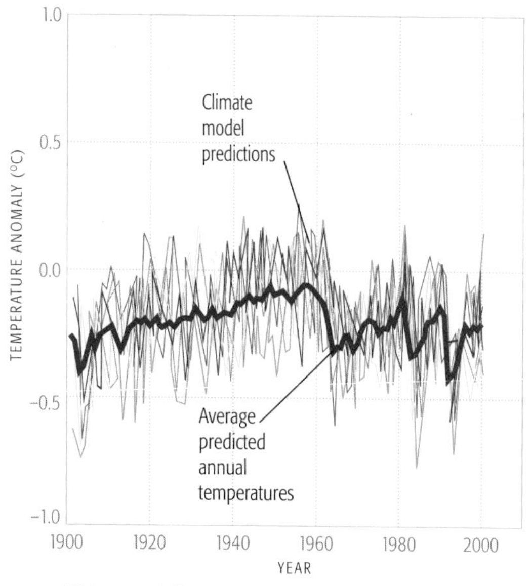

Thirteen different climate models indicate which portion of the annual average temperature variations over the last century can be attributed to natural forces alone.

2 Comparison of the average of the model results in graph 1 to actual observations

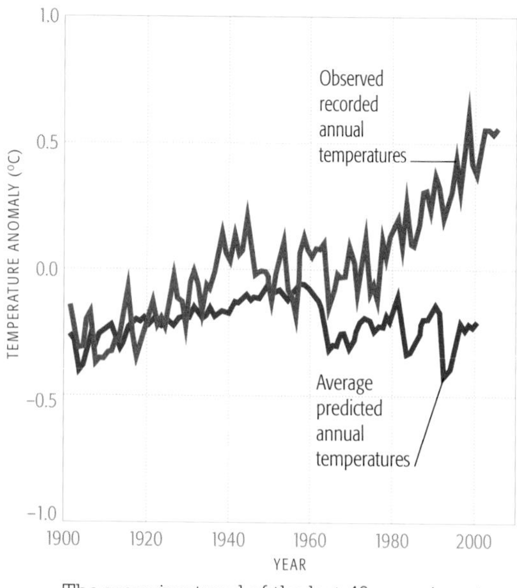

The warming trend of the last 40 years is not reflected in the models that take into account the impacts of natural forces alone.

Mount Pinatubo June 22, 1991
Pinatubo began errupting in the Philippines on June 12, 1991. It caused a subsequent 0.5°C cooling of the atmosphere.

3 Predicted temperature trends from models taking into account the impacts of both natural and human forces

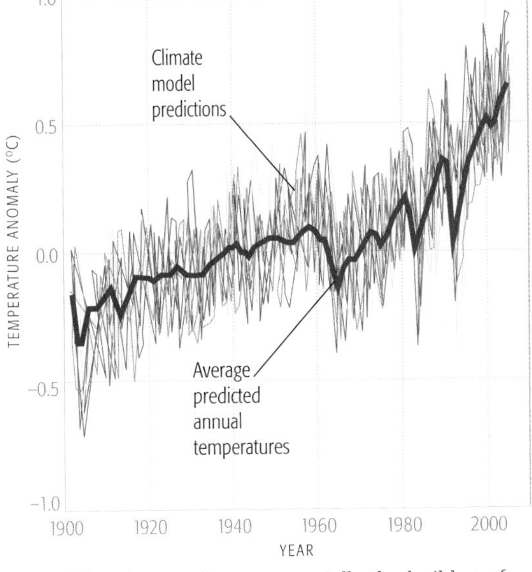

When human forces, especially the buildup of atmospheric greenhouse gases and industrial aerosols, are applied to these same climate models, the models shift to predict a gradual warming over the last several decades.

4 Comparison of the average of the model results in graph 3 to actual observations

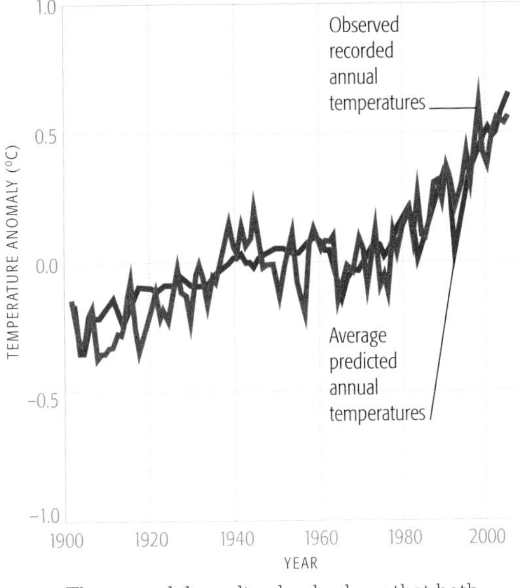

These model results clearly show that both natural and human forces impact climate.

Regional vs global trends

An important development in the latest IPCC report is that it is now possible to tie temperature changes over the oceans, bodies of land, and at the regional scale of individual continents to human activity. When scientists compare observed and modeled temperature changes at these regional scales, they find that the warming trends observed for individual continents, such as North America, Europe, and Africa, cannot be explained by natural factors like volcanoes and changes in solar output. As we have seen on a global scale (see p.68) only when the models include the human, or "anthropogenic," component—warming due to increased greenhouse gas concentrations and the more minor cooling impact of industrial aerosols—can they explain observed regional warming trends (see graphs at right). This indicates that human influences are now having a detectable impact on temperature changes measured in individual regions.

Temperature changes, however, are just one of the many regional impacts of climate change. Influences on other climate phenomena, such as drought and rainfall, may represent even more significant human impacts (see p.89).

Blue planet
This view of the Earth was taken from the space shuttle *Endeavour*.

GLOBAL TRENDS: LAND VS OCEAN

These graphs compare actual observations and model results both with and without human factors included. Only the models that take into account both human and natural factors make predictions that look like actual data trends.

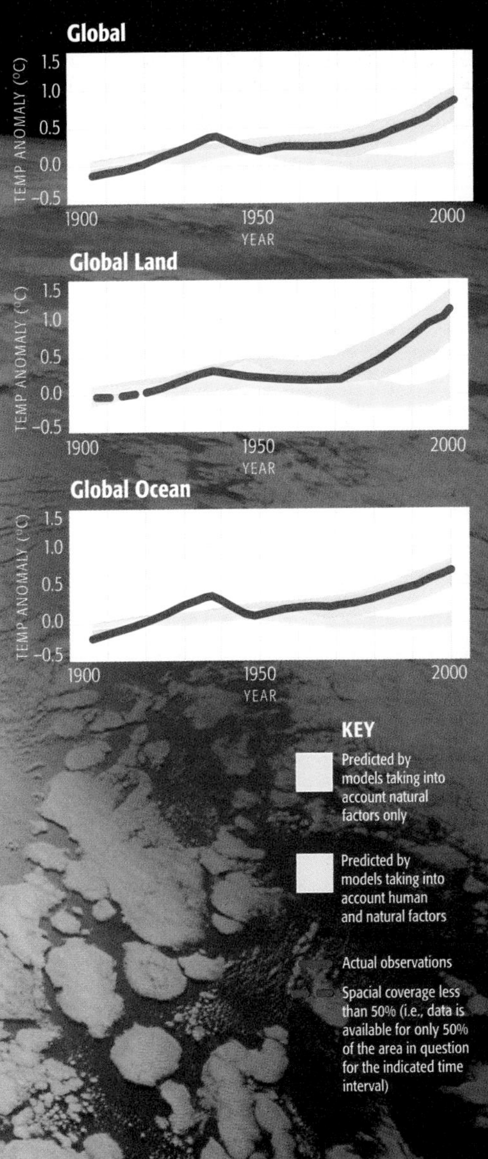

REGIONAL CONTINENTAL TRENDS

These graphs compare actual observations and model results both with and without human factors included. As on the global scale, only models taking into account both human and natural factors make predictions that correspond to actual data.

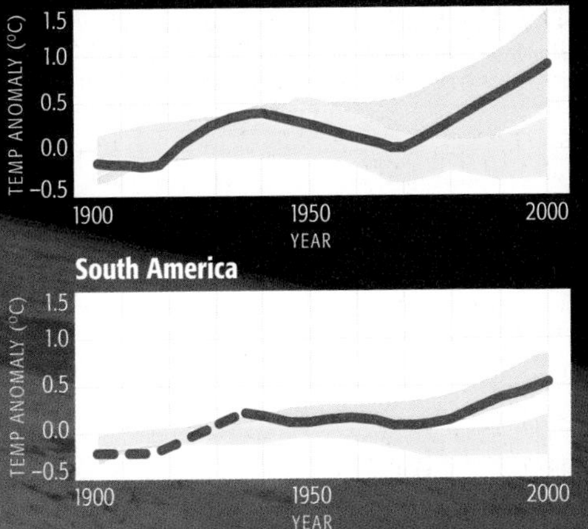

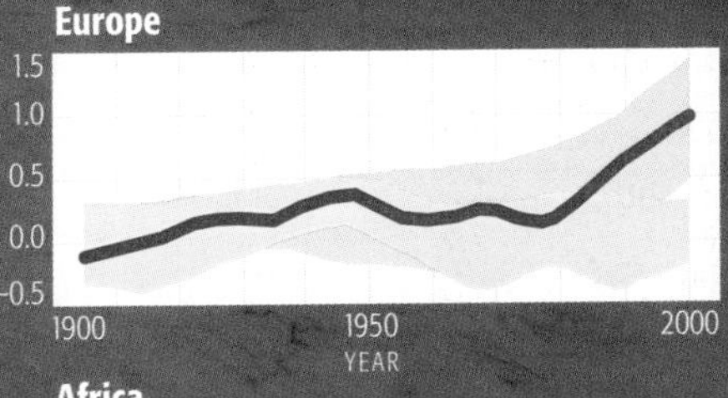

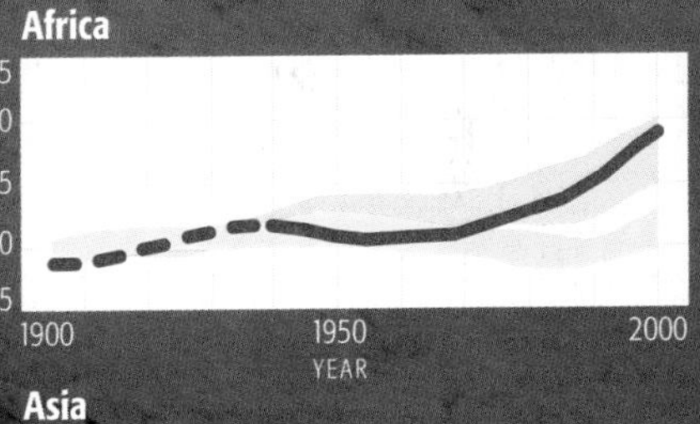

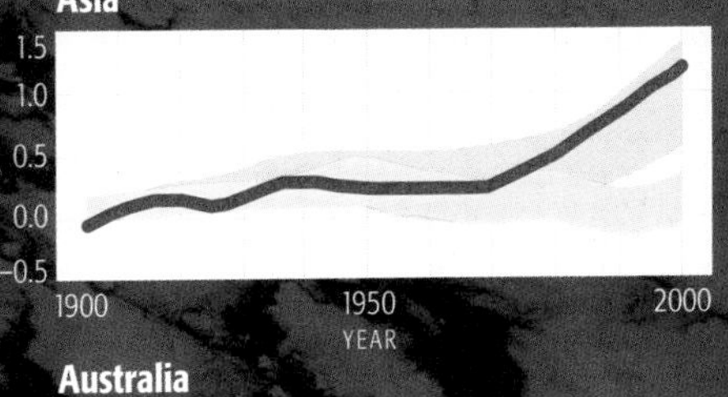

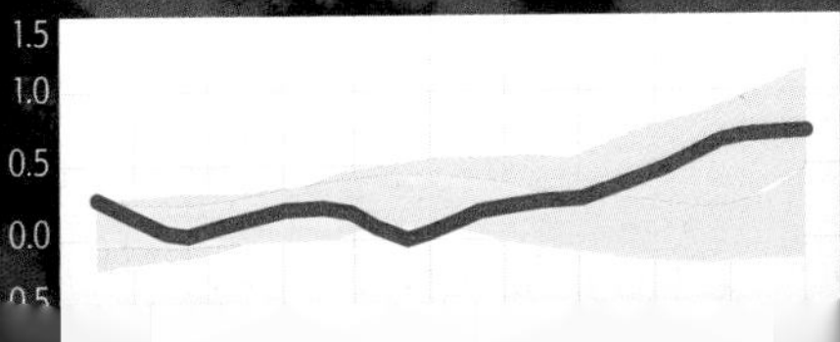

"Fingerprints" distinguish human and natural impacts on climate

Not all factors that affect climate have the same pattern of influence on temperature. One way to distinguish natural and human sources of modern climate change is to look for particular "fingerprints"—that is, specific spatial patterns of change that help to identify the likely underlying factor associated with that change.

Natural impacts
Two natural factors have influenced climate change over the past century:
- Changes in solar output
- Explosive volcanic activity

Human impacts
Major human impacts on climate include the greenhouse gas emissions resulting from:
- Fossil-fuel burning (primary impact)
- Industrial aerosols (secondary impact) (see p.18)

We have seen that on both the global (see p.68) and regional scale (see p.70) climate model simulations which include all of the natural and human factors together can do a fairly good job in reproducing the observed pattern of surface warming. By contrast, simulations that include only the natural factors are unable to reproduce nearly any of the observed surface temperature changes. This observation holds when we compare simulations at the detailed spatial level, as seen in the global maps to the right.

Climate model simulations that include only natural factors reproduce hardly any of the observed surface temperature changes.

Natural climate model surface temperature calculation 1979–2005

Human and natural climate model surface temperature calculation 1979–2005

Actual recorded surface temperatures 1979–2005

WARMING PATTERNS

The pattern of warming over the past few decades is not reflected in the models that only take into account the impacts of natural forces alone (top map). Instead, actual observations (bottom map) correspond closely with model predictions that take into account the impacts of both natural and human forces (middle map).

Surface temperature key

Coal fire

Coal-fired power stations are among the primary sources of industrial greenhouse gas emissions.

Atmospheric fingerprints

Another fingerprint is the pattern of expected atmospheric temperature changes. Different factors, natural and human-generated, have different effects on the different layers of the atmosphere (see p.38). Increases in solar output are predicted to warm essentially the entire atmosphere from top to bottom. Volcanoes cool the lower atmosphere (the troposphere) slightly, and warm the mid-level atmosphere (the stratosphere).

By contrast, human-generated increases in greenhouse gases are predicted to warm the lower atmosphere (the troposphere) substantially, at the expense of cooling the upper atmosphere (the stratosphere and above).

Human-generated greenhouse gas concentration increases are thought to be the primary cause of atmospheric temperature changes over the past century. Not surprisingly then, the

Solar effect on atmospheric temperature 1890–1999

Volcanic effect on atmospheric temperature 1890–1999

Human-generated greenhouse gas effect on atmospheric temperature 1890–1999

ATMOSPHERIC TEMPERATURE CHANGE

The model-predicted pattern of atmospheric temperature change taking into account the impacts of both natural and human influences (far right) looks much like the greenhouse gas pattern alone (second from right). This is because human-generated increases in greenhouse gas concentrations are believed to have dominated atmospheric temperature changes over the past century.

Atmospheric temperature key

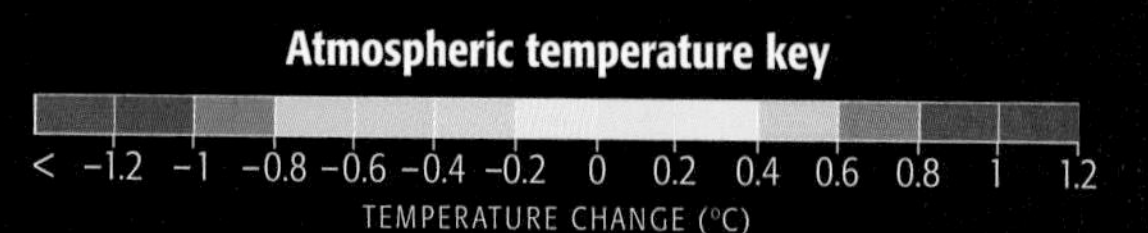

Atmospheric layers are not drawn to scale; height has been exaggerated in order to show color variations as clearly as possible.

expected pattern of atmospheric temperature change due to all impacts combined (natural and human) looks much like the greenhouse gas pattern alone. Indeed, we see that this pattern matches the observed pattern of temperature change in recent decades (see p.38).

Combined effect of human and natural forces on atmospheric temperature 1890–1999

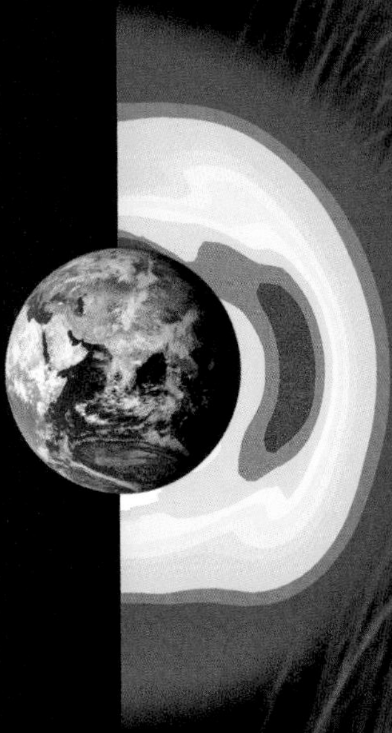

Natural fireworks
Stromboli, in Italy, is one of the most active volcanoes on Earth.

Part 2 Climate Change Projections

Projections of how Earth's climate will change are uncertain. They depend on both the unknown future trajectory of greenhouse emissions, and the uncertain response of the climate to these emissions. Nonetheless, researchers can draw certain conclusions given best-guess scenarios of fossil-fuel burning and the average predictions of theoretical climate models. Scientists can project, for example, that for "middle of the road" emissions scenarios, the globe is likely to warm by several more degrees Celsius by the end of the 21st century. This warming is likely to be associated with a dramatic decrease in Arctic sea-ice, an acceleration of sea level rise, and increased drought, flooding, and extreme weather for many regions of Earth.

How sensitive is the climate? Modern evidence

To determine the potential magnitude of future global warming, scientists find it useful to estimate something they call "climate sensitivity."

- **Climate sensitivity** defines the amount of warming (in degrees Celsius) that we can expect to occur when there is a change in the factors that control climate. It is a way of placing a numerical value on how much our planet will warm in response to future increases in greenhouse gas emissions. Climate sensitivity is typically expressed in terms of the expected surface warming that will occur in response to a doubling of atmospheric CO_2 levels from their pre-industrial level of roughly 280 parts per million (ppm) by volume.

- **Equilibrium climate sensitivity** takes into account the fact that the full amount of warming in response to an increase in greenhouse gas concentrations may not be realized for many decades, due to sluggish ocean warming. In plain language, this means that if we say equilibrium climate sensitivity is 3°C, we mean that Earth will eventually warm by 3°C if CO_2 levels reach 560 ppm by volume. At current rates of fossil-fuel burning, this doubling of CO_2 levels is expected to occur mid-way through the 21st century. However, the resulting warming may not fully be experienced until at least 2100.

So how do scientists estimate climate sensitivity?

Using climate models, scientists compare observed temperature changes (from the instrumental record of the past 150 years) with simulations of temperature changes over this same time frame (see p.68). In order to determine the actual climate sensitivity, certain types of climate models are tuned to different climate sensitivity values. Scientists then determine which of these sensitivity values best match the observed temperature changes. By looking at the various climate sensitivities that fit reasonably well with the actual temperature record, the scientists can quantify the uncertainty of the estimated climate sensitivity values. This range is quite large if only the relatively short instrumental record is used.

SIMULATED VS ACTUAL SURFACE TEMPERATURE CHANGES

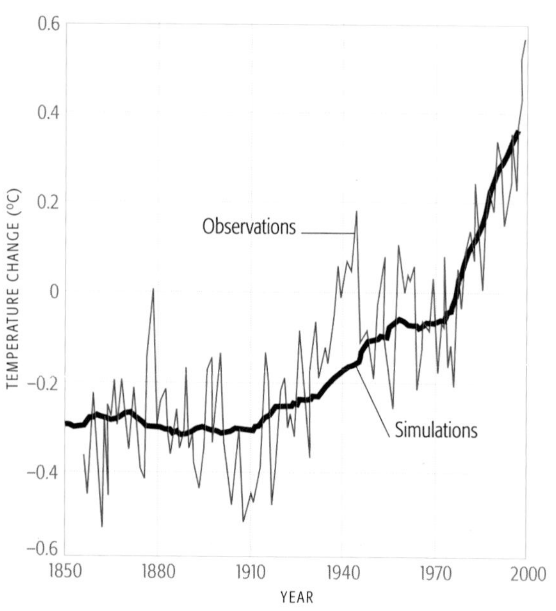

A climate sensitivity of 2.7°C was assumed in these simulations, which took into account both human and natural factors over the 150 years for which surface temperature observations exist.

Deep ocean temperature

Over the past few decades, it has also been possible to make use of temperature measurements from the deep oceans. While such data are useful, they are limited by an even shorter available record.

SIMULATED VS ACTUAL DEEP-OCEAN TEMPERATURE CHANGES

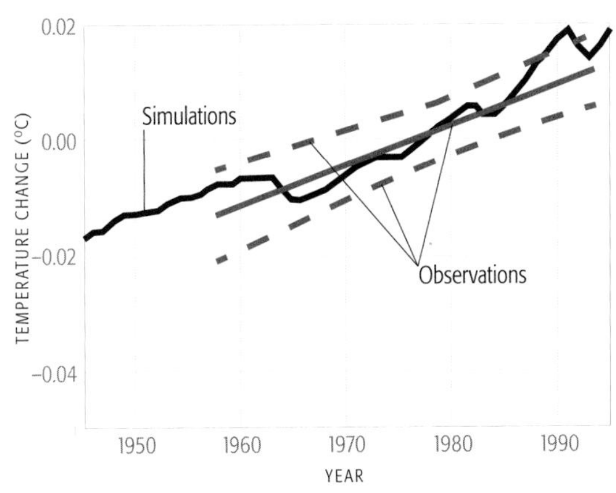

A climate sensitivity of 2.9°C was assumed in these simulations, which took into account both human and natural factors over the 50 years for which deep-ocean temperature observations exist.

Do we need more data?

Overall, these simulations yield an estimate of equilibrium climate sensitivity of roughly 3°C. In other words, it is estimated that Earth's surface will warm by 3°C if CO_2 concentrations double. Uncertainties, however, are large.

Modern instrumental observations could be consistent with a climate sensitivity anywhere from 1.5°C to 9°C.

With such a wide temperature range, the effect of climate change could be anything from essentially negligible to wholly catastrophic. This uncertainty is inevitable when we only have access to a short (150 years or less) record. There are so many different natural and human factors that are simultaneously at play, and each has impacts that are individually uncertain. For this reason, scientists turn to other longer-term sources of information. In the following pages, we show how this is done.

Floating monitors
Weather buoys gather data to keep scientists informed of climate changes on the surface of the oceans.

How sensitive is the climate? Evidence from past centuries

Another way to estimate climate sensitivity is to study responses to changes in the natural factors governing climate in previous centuries. Using information from climate proxy data (see p.40), such as tree rings and ice cores, scientists estimate how the average temperature of the northern hemisphere varied during the past millennium. They also estimate how the natural factors influencing Earth's climate changed over this time frame, and then compare the two sets of data. Of course, it is important to note that all these estimates come with substantial uncertainties.

Information about how the factors that govern climate have changed takes many forms. Sunspot records are available from the early 17th century up to modern times. Chemical substances that fall to the surface in snow and become trapped in ice cores can be used to track solar activity even further back in time. Explosive volcanic eruptions can be documented through analyses of the aerosol deposits they left behind in ice cores. The long-term increase in greenhouse gas concentrations since the advent of industrialization is documented in the content of air bubbles trapped within the ice (see p.32).

Consistent sensitivity estimates

The equilibrium climate sensitivity estimate of 2–3°C derived from proxy data is similar to estimates produced using the modern record (see p.78) and geological data (see p.82).

Sunspot record

Measurements of sunspots date back to the early 1600s, when they were first recorded with telescopes by European astronomers, such as Galileo. Modern satellite measurements demonstrate that the sun is slightly brighter in years with high sunspot counts.

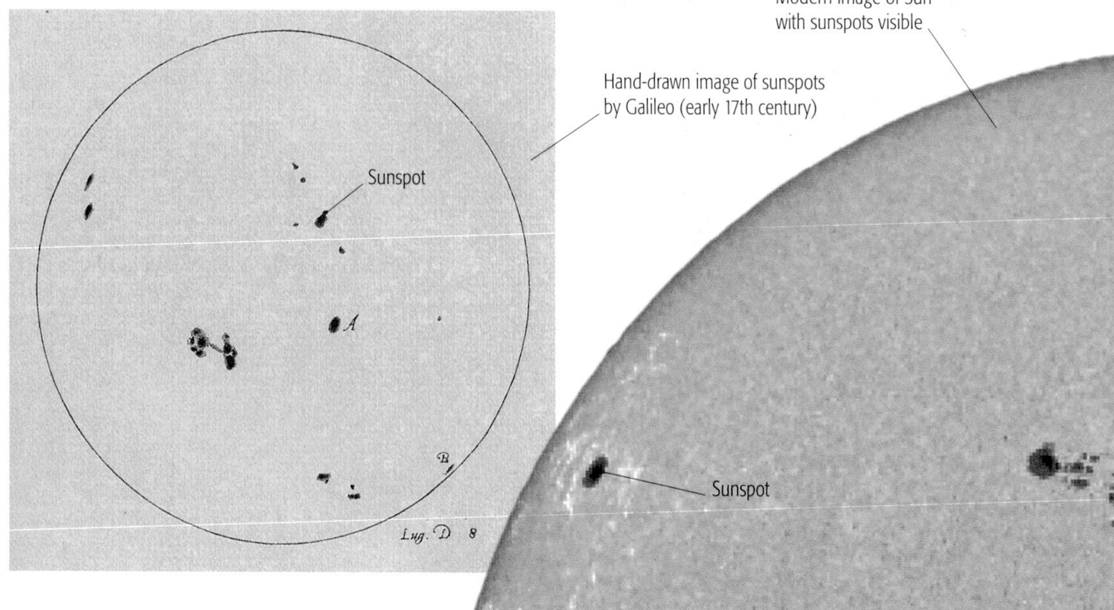

NORTHERN HEMISPHERE TEMPERATURE CHANGES OVER THE PAST SEVEN CENTURIES: SIMULATED VS ESTIMATES FROM PROXY DATA

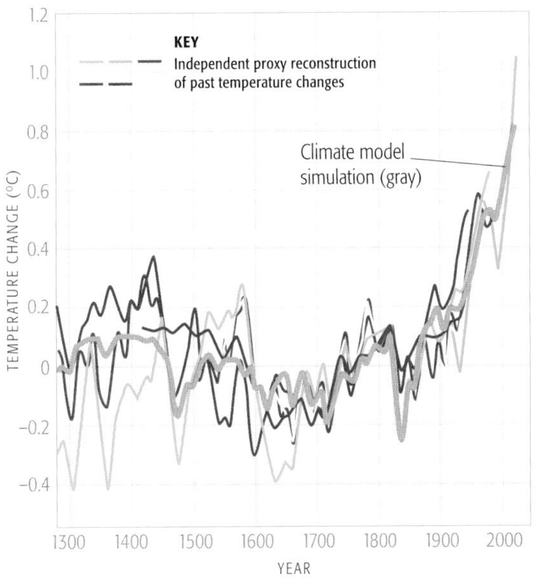

Climate scientists compare model predictions with estimated changes in average temperatures in the northern hemisphere derived from proxy data. The proxy temperature estimates match the model simulations well when the assumed equilibrium climate sensitivity is 2–3°C, meaning that a doubling of atmospheric CO_2 concentrations will lead to a roughly 2–3°C warming of the globe.

ESTIMATES OF NATURAL AND HUMAN IMPACTS ON CLIMATE OVER THE PAST 1000 YEARS

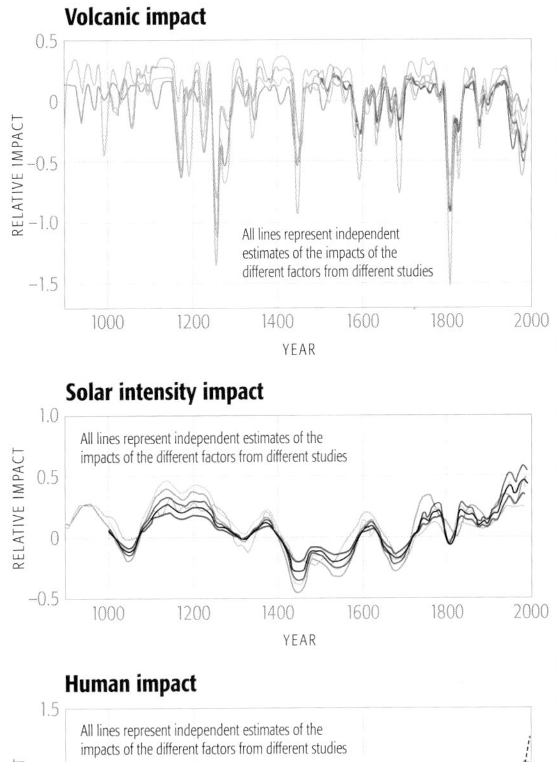

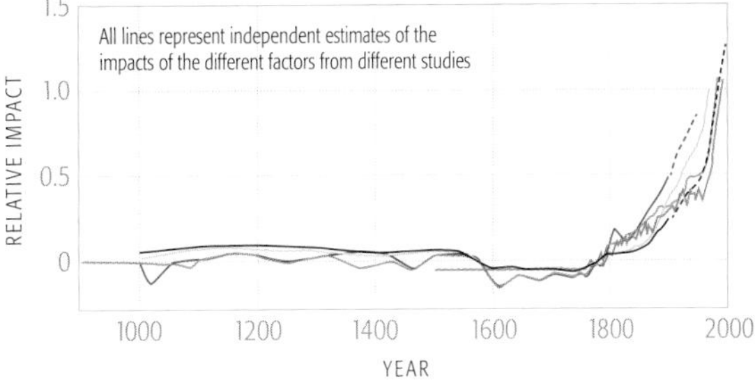

Scientists drive climate models with the estimated impacts of both natural and human forces. Individual volcanic eruptions have a significant short-term impact, but collectively volcanoes can drive long-term changes when their frequency and magnitude change from one century to the next. Fluctuations in solar output take place on centennial-long timescales. Human-caused greenhouse gases have ramped up dramatically over the past two centuries.

How sensitive is the climate? Evidence from deep time

As we've seen, studies of climate change over the last few centuries can provide us with reliable estimates of climate sensitivity. These sensitivities correspond to changes in atmospheric CO_2, ranging from the pre-industrial level of 280 ppm to the 2008 value of 386 ppm. While significant, this range doesn't include the known glacial–interglacial variations in atmospheric CO_2 over the last 650,000 years: at the peak of the glacial periods, atmospheric CO_2 dipped to 180 ppm. The range also comes well short of possible future increases in atmospheric CO_2, which are predicted to reach nearly 2000 ppm. So how do we determine how climate will respond to the significantly elevated levels of atmospheric carbon dioxide anticipated for the future? We need to look to the ancient past for clues.

Clues from deep time

Geologists estimate that ancient atmospheres contained as much as 2000 ppm of CO_2, and even more (see p.40). Therefore studies of ancient climates can provide important information on climate sensitivity for much larger CO_2 ranges.

A turbulent past

For the last 2 million years, Earth has been swinging in and out of glacial conditions, driven by subtle changes in Earth's orbit around the Sun that are amplified by feedbacks in the carbon cycle and climate system. Data from ice cores demonstrate that fluctuations in CO_2 and temperature have gone hand in hand for at least the last 400,000 years. Feedback loops in the carbon cycle make the question of whether CO_2 is driving climate changes or vice versa virtually impossible to answer. Nevertheless, computer models only simulate the observed cooling when input with low atmospheric CO_2 levels.

Then and now

To learn more about how climate responds to different CO_2 levels, let's step back in time to the height of the last ice age (the "Last Glacial Maximum," or LGM) 21,000 years ago. With much less CO_2 in the atmosphere, the world was then quite a

During the last ice age, sea level was 120 m lower

CHANGE IN SEA SURFACE TEMPERATURE NOW COMPARED TO 21,000 YEARS AGO

Sea surface temperature differences between the Last Glacial Maximum, 21,000 years ago, and today, show general cooling. Mid-to-high latitudes experienced more intense cooling (dark blue), especially near the ice sheets (shown in white).

different place. The sea level was 120 m lower, because evaporated seawater had fallen as snow and formed the vast ice sheets of the northern hemisphere. A stroll to the beach from Atlantic City, New Jersey, would have taken days, since the shoreline was 80 km east of where it is today. Based on ice-core gas analyses (see p.32), we know that the atmosphere's CO_2 content was about 50% of what it is now. There were a number of other differences between the LGM and today:

- Atmospheric methane was about one-fifth and nitrous oxide was about two-thirds of what they are today.
- Vast ice sheets covered much of Canada, the northernmost US, Scandinavia, and northern Europe. These ice sheets were considerably more reflective than the surfaces they replaced. This accounts for half of the cooling, since the ice sheets were reflecting heat rather than absorbing it.
- Earth's orbital configuration was different than it is today (see p.62). Because of this, the amount of summer sunshine at high northern latitudes was reduced, so snow from the winter survived the summer and additional ice accumulated.

What can the Last Glacial Maximum teach us about tomorrow's climate?

CO_2, the LGM, and today

Climate scientists have taken on the challenge of assessing the observed climate of the Last Glacial Maximum. They want to know if the way climate behaved then in response to changes in CO_2 can help us understand how it will behave in the future. The compiled data from the previous page indicate that the global average temperature at the LGM was 5°C cooler than it is today and atmospheric CO_2 levels were much lower as well (180 ppm). The equilibrium climate sensitivity estimate for the LGM is 2.3–3.7°C, satisfyingly close to our other estimates (see p.80). This tells us that data from the LGM confirm our predictions for how the climate will respond to a doubling of atmospheric CO_2.

Ancient data

How will climate respond to even higher levels of CO_2 than those experienced in the glacial–interglacial fluctuations? To answer that question, we must venture much further back into Earth's history. Because ice cores do not go back this far, there is no direct measure of atmospheric composition. So geologists have developed a variety of proxy methods to study atmospheric CO_2 levels. Each tells a somewhat different story, but the overall trends, as shown in the graph below, are consistent. Atmospheric CO_2 levels were high 400 million years ago, then fell, reaching a minimum 300 million years ago. After that, levels rose and fell, but reached another maximum about 175 million

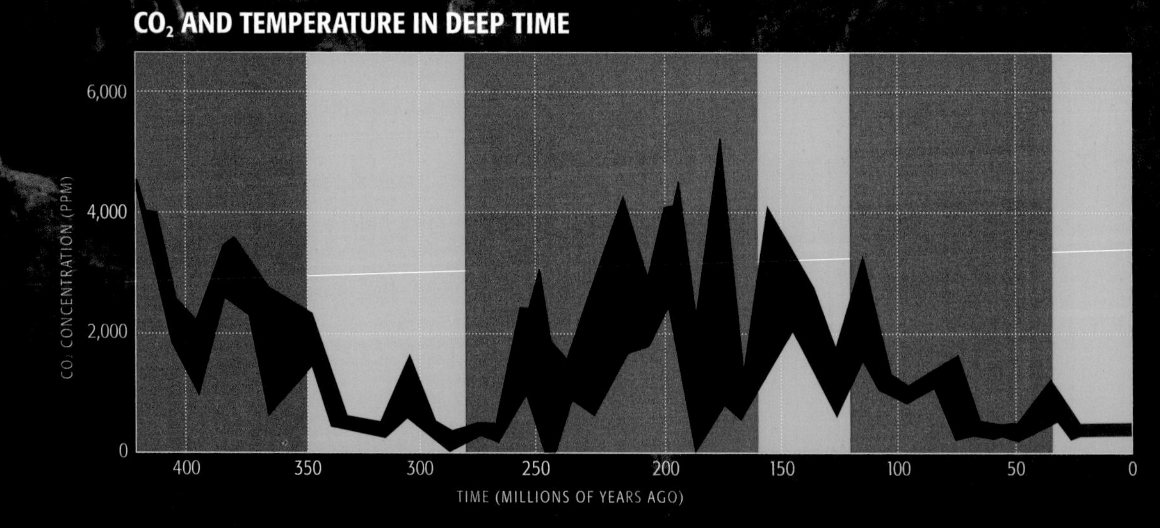

Reconstructions of atmospheric CO_2 levels and global climate from proxy evidence demonstrate the relationship between the two over the course of geological time.

KEY

Warmer than today

As cold, or cooler than today

years ago in the late Triassic. They stayed relatively high through the Mesozoic (252–65 million years ago). This was the age of the dinosaurs, when crocodile-like reptiles ventured above the Arctic Circle. Since the late Triassic peak, atmospheric CO_2 levels have generally fallen, reaching another minimum very close to the present day. In the graph below left, you can see that climates (red for warmer than today; blue for as cold, if not colder, than today) generally corresponded to these CO_2 fluctuations. The one exception was the relatively cool interval from 160 to 130 million years ago, which remains unexplained. When combined with the actual paleotemperature estimates, these proxy CO_2 data provide a specific estimate of equilibrium climate sensitivity of 2–5°C for each doubling of atmospheric CO_2, which is entirely consistent with data from the LGM and with the predictions from state-of-the-art climate models. Most importantly, this analysis precludes a weak equilibrium climate sensitivity (less than 1.5°C per doubling). This confirms the general notion that substantial greenhouse warming is the expected consequence of a buildup of atmospheric CO_2.

A billion years of Earth history
Strata in the Grand Canyon contain evidence of large swings in climate that geologists can relate to corresponding fluctuations in atmospheric CO_2 levels.

Fossil-fuel emissions scenarios
Predicting the possibilities

Lurking underneath all our predictions about future climate change is a vexing uncertainty: how will human consumption of fossil fuels and land use practices evolve over the next decades and centuries? The driving forces for consumption are highly complex, involving population growth and per-capita energy demands. These factors are, in turn, closely linked to economic growth and technological advances that can both accelerate consumption and also shift it to other climate-neutral sources.

Possible scenarios

In an attempt to learn more about an uncertain future, climate researchers create and evaluate a range of scenarios for greenhouse gas emissions. This exercise helps them to determine the scope of consequences for a variety of possible future fuel-use scenarios. Initially, these were divided into "business as usual" scenarios, which assume ever-increasing rates of fossil-fuel use, and "conservation" scenarios, which assume some future reduction of use. For the most recent IPCC assessment, experts from around the world developed four basic "storylines", each representing a group

POSSIBLE CO_2 EMISSIONS SENARIOS FOR THE FUTURE

A1 storyline – one global family

A1F1
A1B
A1T
GLOBAL CO_2 EMISSIONS (GIGATONS PER YEAR)
40
30
20
10
0
1990 2010 2030 2050 2070 2090
YEAR

- Substantial reduction of regional differences in per-capita income
- Rapid economic growth
- Peak population mid-21st century, declining thereafter
- Rapid introduction of new, more efficient technology

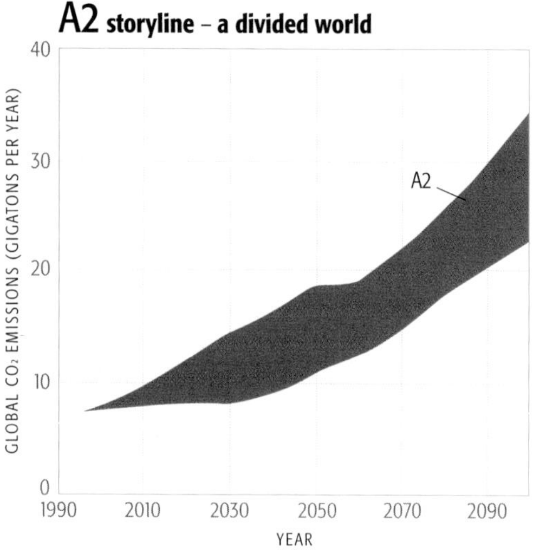

- Emphasis on national identities and local and regional solutions to environmental protection and social equity issues
- Slow per-capita economic growth and technological advancement
- Continuously increasing world population

of emissions scenarios for the future. The four storylines were designated A1, A2, B1, and B2. To take into account how alternative energy technology developments might affect the climate projections, scientists also refer to three groups of scenarios within the A1 storyline: fossil-fuel intensive (A1FI), non-fossil fuel intensive (A1T), and a scenario that assumes a balanced use of both fossil and non-fossil fuels (A1B). The A1B scenario is a "middle of the road" scenario often used as a basis of comparison in the IPCC report. Each scenario envisions a different future path for Earth and its citizens.

Which one will it be?

Future emissions differ quite dramatically between storylines. The largest growth and cumulative release of CO_2 is associated with the A1FI fossil-fuel-intensive scenarios, while the smallest is associated with the B1 scenarios.

The degree of overlap among these emissions scenarios indicates that very different socioeconomic factors can lead to similar levels of CO_2 emissions in the future. And the spread of values for any given storyline reveals that similar socioeconomic factors can lead to quite different atmospheric CO_2 levels.

The IPCC scientists made no attempt to estimate likelihoods for any of these possible scenarios actually occurring; the uncertainties are simply too large. However, these projections do give climate modelers and social scientists a reasonable range of emissions scenario options with which to work.

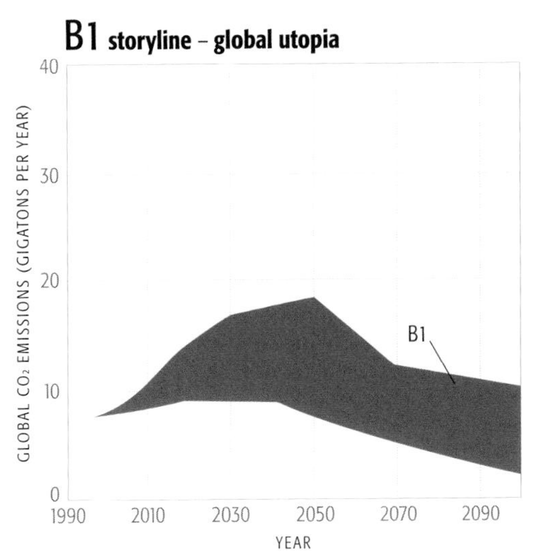

- Emphasis on global solutions to sustainability and environmental protection
- Rapid change to information and service economy
- Peak population mid-21st century, declining thereafter, as in A1
- Reduction in intensity of demand for materials
- Introduction of clean and efficient energy technologies

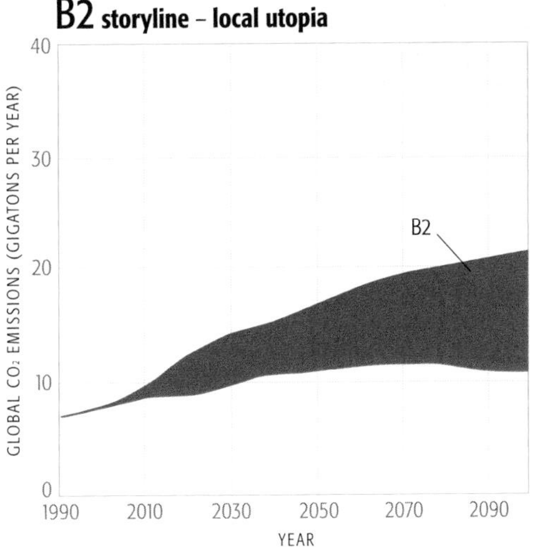

- Continuously increasing world population, slower growth than in A2
- Intermediate levels of economic development
- Slower development of new energy technologies than B1 and A1
- Emphasis on local and regional rather than global solutions to environmental protection and social equity issues

The next century
How will the climate change?

Scientists have come up with a range of possible trajectories for future climate change using climate models (see p.64). The spread of these trajectories is due to differing possible future greenhouse emissions scenarios (see p.86), as well as the variances among individual climate models, which differ in their climate sensitivity (see p.78). Typically, the results from several different climate models are averaged to yield a single trajectory that corresponds to a particular emissions scenario.

Temperature changes

The predicted increase in global average temperature from 2000 to 2100 is roughly:

- 1–3°C for the most aggressive emissions scenario (B1 in figures below).
- 1.5–4.5°C for the "middle of the road" scenario (A1B in figures below).
- 2.5–6.5°C for the least aggressive scenario (A1FI in figure below right).

These model projections do not take into account some possible positive feedbacks (see p.94) that could further exacerbate global warming. In certain regions, moreover, warming may be considerably greater than the average predicted for the globe as a whole (see p.92).

It is worthwhile noting that only the most conservation-minded scenarios are likely to avoid warming in excess of 2°C. This is the benchmark rise that is often cited as constituting dangerous human interference with the climate (see p.108).

ESTIMATED TEMPERATURE TRAJECTORIES FOR VARIOUS EMISSIONS SCENARIOS

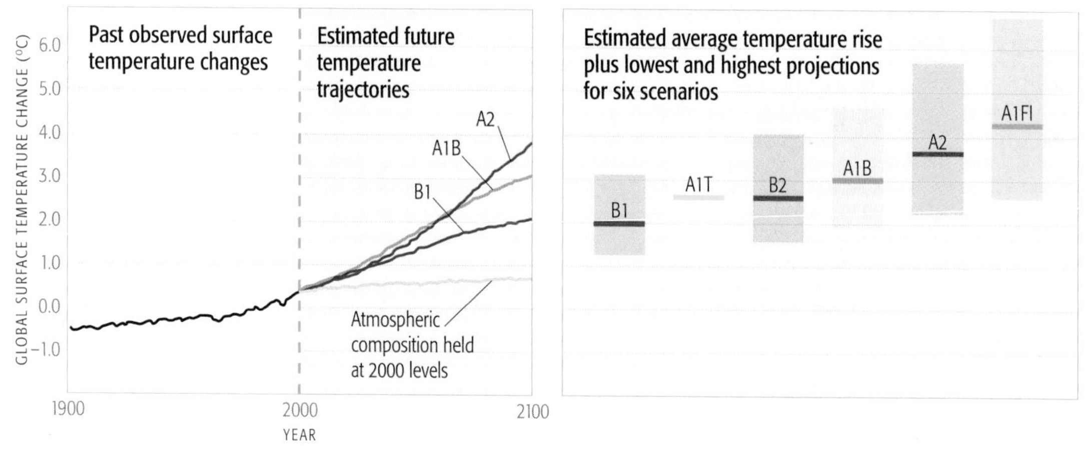

The graph at left shows the estimated temperature trajectories for various emissions scenarios, as predicted for the various scenarios by the 23 different state-of-the-art climate model simulations used in the most recent IPCC report. The bold-colored straight lines in the graph at right show the averages for various scenarios, while the surrounding lighter bars show the range between the lowest and highest predictions.

HADLEY CIRCULATION PATTERN

In the Hadley circulation pattern, warm moist air tends to rise, cool, and produce rain near the equator. Depleted of its moisture, it eventually sinks as dry air in the subtropics.

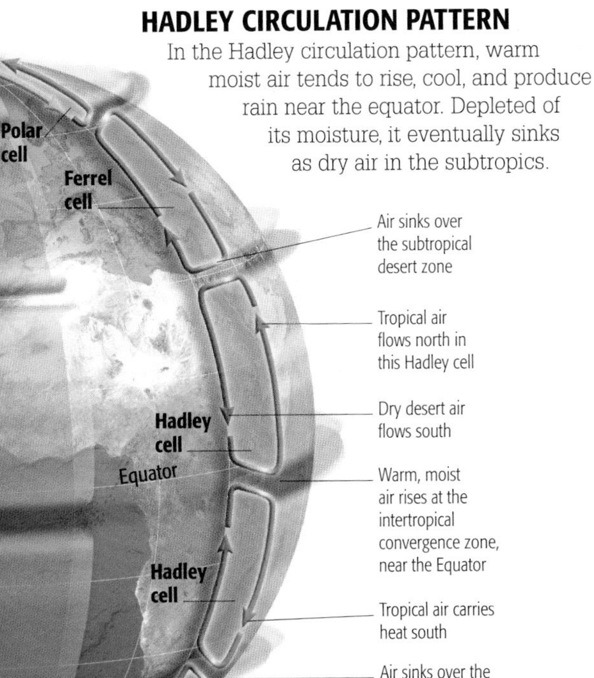

Precipitation changes

Perhaps of more profound importance than temperature changes are the projected changes in precipitation.

The projected poleward shift in the jet streams of both hemispheres may cause:

- Increased winter precipitation in polar and subpolar regions
- Decreased precipitation in middle latitudes

Poleward expansion of the tropical Hadley circulation pattern will cause:

- Decreased precipitation in the subtropics

A warmer atmosphere will cause:

- Increased precipitation near the equator

PRECIPITATION PROJECTIONS

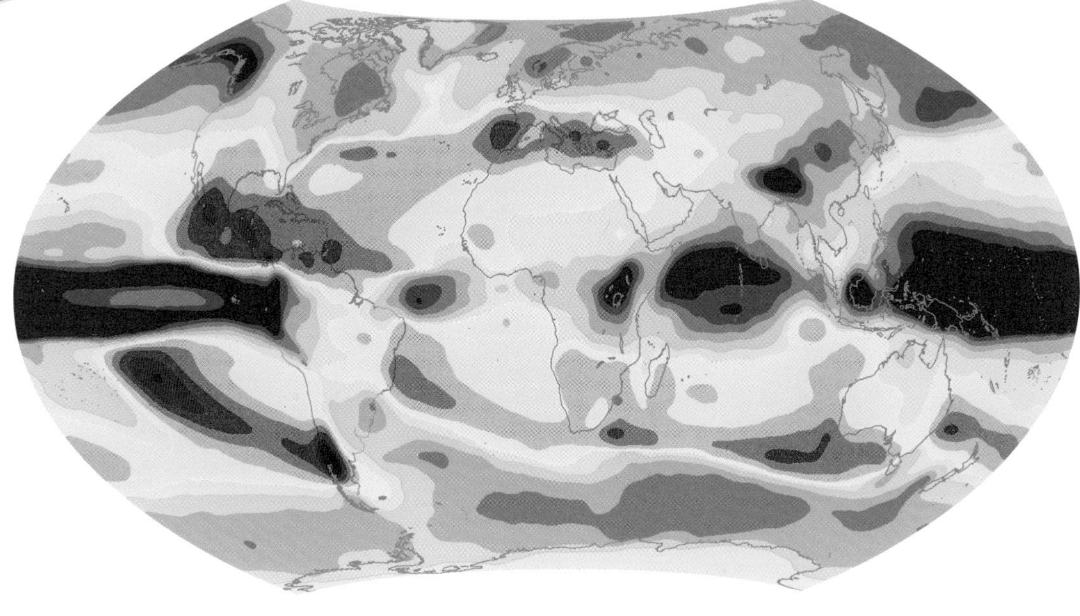

Precipitation pattern changes (relative to 1980–1999) projected to occur by 2100 in response to the so-called "middle of the road" emissions scenario. Note that there is increased precipitation prodicted near the equator and in subpolar regions, while subtropical and mid-latitude regions will likely become drier.

More drought, more floods

The combination of decreased summer precipitation and increased evaporation due to warming surface temperatures is predicted to lead to a greater tendency for drought in many regions. The more vigorous cycling of water through the atmosphere favored by a warming globe will lead to greater rates of both evaporation and precipitation. Consequently, more frequent intense rainfall events and flooding can be expected for many regions as well. Other likely impacts of climate change over the next century include increases in extreme weather phenomena (see p.100), and rising sea levels due to melting ice and the warming of the oceans (see p.98).

El Niño Southern Oscillation (ENSO)

The El Niño-Southern Oscillation (ENSO) is a natural irregular oscillation of the climate system, involving inter-related changes in ocean surface temperatures, ocean currents, and winds across the equatorial Pacific. This phenomenon alternates every few years between so-called "El Niño" and "La Niña" events, which influence weather patterns across the globe.

EL NIÑO EVENT

During El Niño events, the so-called "trade winds" in the eastern and central tropical Pacific weaken or even disappear, there is little or no upwelling of cold sub-surface ocean water in the eastern equatorial Pacific, and warm water spreads out over much of the tropical Pacific ocean surface.

LA NIÑA EVENT

During La Niña events, the trade winds in the eastern and central tropical Pacific are stronger than usual, and there is strong upwelling of cold, deep water in the eastern and central equatorial Pacific.

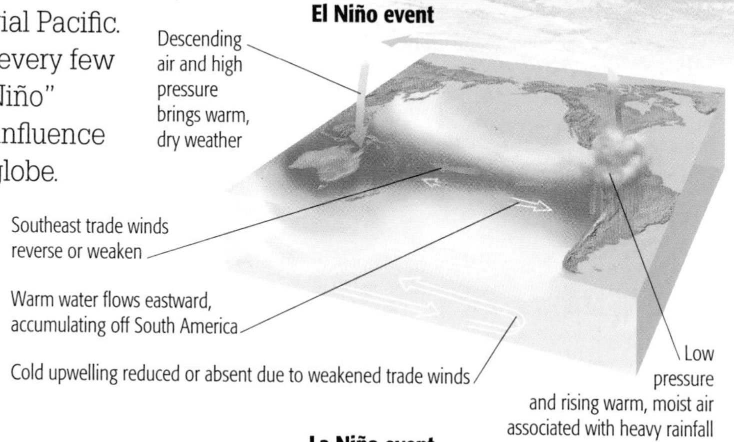

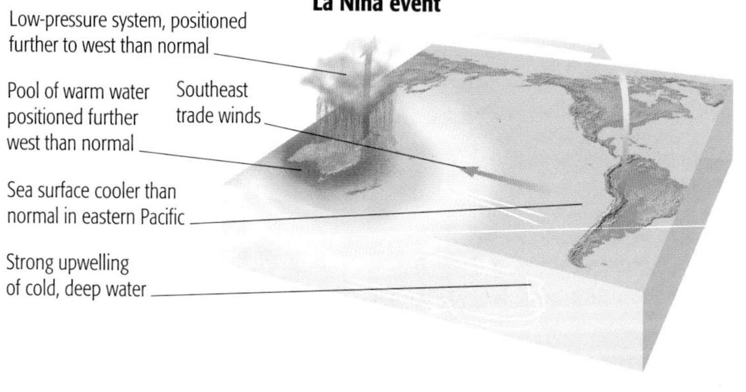

LARGE-SCALE IMPACTS OF EL NIÑO (NORTHERN HEMISPHERE WINTER)

El Niño events influence global patterns of temperature and rainfall. The effects of La Niña events are roughly opposite to those shown here for El Niño events

Dry | Dry and warm | Warm | Wet | Wet and cool | Wet and warm

Indonesian rainstorm
A man rides his motorcycle with his son through a flooded street in the city of Tangerang, west of Jakarta. On February 13, 2003, heavy rain pelted Jakarta and surrounding cities for five hours, triggering floods in several parts of the area around the capital.

ENSO VARIABILITY

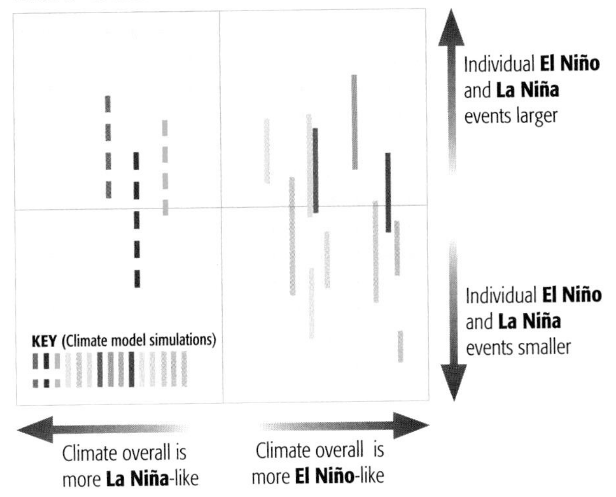

Most of the state-of-the-art climate model simulations used in the most recent IPCC report predict a more El Niño-like pattern in response to climate change, but some models predict an opposite, La Niña-like pattern. Models are nearly equally split as to whether the year-to-year ENSO variability is likely to increase or decrease in magnitude.

Uncertain ENSO

Precise regional future climate change predictions are hampered by uncertainties in how global wind patterns and ocean currents will change. Models don't yet agree on the basic question of whether the climate will become more or less El Niño-like in response to human impacts on climate. Since ENSO is such an important influence on regional patterns of precipitation and temperature, such uncertainties translate to an uncertainty about the patterns of regional climate change themselves. If El Niño events become more frequent, then winter precipitation will increase in regions such as the desert southwest of the US, offsetting any trend toward increased drought in the region (see p.48). More El Niño events would also favor a worsening of drought in southern Africa and other regions.

The geographical pattern of future warming

Melting ice field
The extent of summer sea-ice in the Arctic Ocean has greatly diminished in recent decades.

The pattern of projected warming over the next century is far from uniform. The greatest warming will take place over the polar latitudes of the northern hemisphere, due to the positive feedbacks associated with melting sea-ice. Greater warming is projected for land masses than for ocean surfaces, due mostly to the fact that water tends to warm or cool more slowly than land. Accordingly, there is greater warming in the northern hemisphere, which has a higher proportion of land mass, than the ocean-dominated southern hemisphere. Some of the regional variation in warming is due to changes in wind patterns and ocean currents that are also produced by the changing climate. Relatively little warming, for example, is projected to take place over an area of the North Atlantic ocean just south of Greenland, because weakening ocean currents (see p.60) and shifts in the pattern of the northern hemisphere jet stream favor a greater tendency for cold-air outbreaks in this region.

MODEL-PROJECTED WARMING FOR 2030 AND 2100

These maps shows projected surface temperature changes relative to the average temperatures during the late 20th century for the A1B "middle of the road" emissions scenario (see p.86).

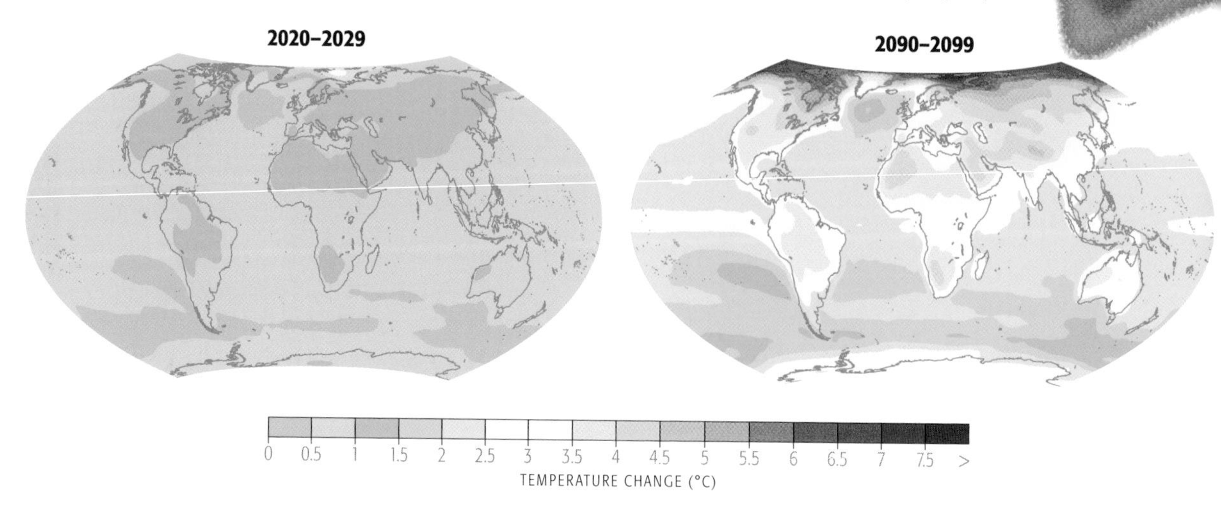

6.5°C WARMER

Breaking down the projected pattern of warming at continental scales, it is clear that North America is likely to see the greatest warming, while South America and Australia are likely to see more modest warming. It should be kept in mind, however, that precise regional temperature projections are limited by uncertainties in how the El Niño/Southern Oscillation phenomenon and other regional atmospheric circulation patterns will be affected by climate change (see p.90).

CONTINENTAL SURFACE TEMPERATURE ANOMALIES: OBSERVATIONS AND PROJECTIONS

The projected future warming in the "middle of the road" emissions scenario (see p.86) for each continent is well beyond the range of temperature changes seen over the past century.

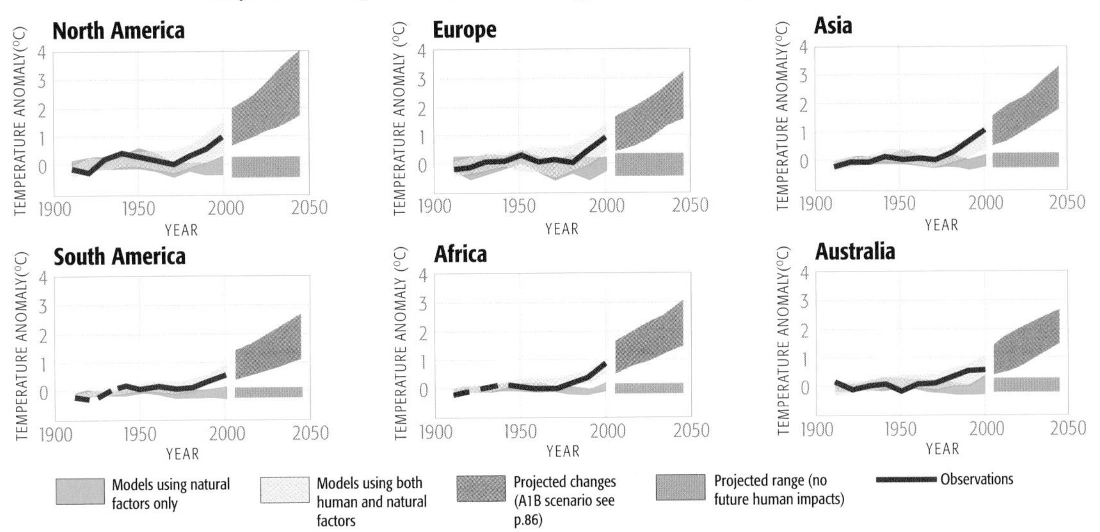

Carbon-cycle feedbacks
Nature's response to CO_2

You might think that all of the carbon dioxide released through fossil-fuel burning and deforestation has simply accumulated in the atmosphere. Yet detailed analysis shows that 45% of the CO_2 we've pumped into the atmosphere since 1959 has "disappeared." Actually, scientists know where it went. Much of the "missing" CO_2 has dissolved into the ocean, and the rest has been stripped from the atmosphere via photosynthesis and incorporated into living "biomass." (Photosynthesis is the process by which plants, and a few other organisms, use energy from sunlight to convert CO_2 into sugar—the "fuel" used by all living things on Earth.) The 55% of the CO_2 that has not "disappeared" but accumulated in the atmosphere is termed the "airborne fraction." In effect, nature has already responded to fossil-fuel burning to a certain degree, and somewhat reduced the human impact on atmospheric composition and climate, but nature has its limits and humans are beginning to push up against them.

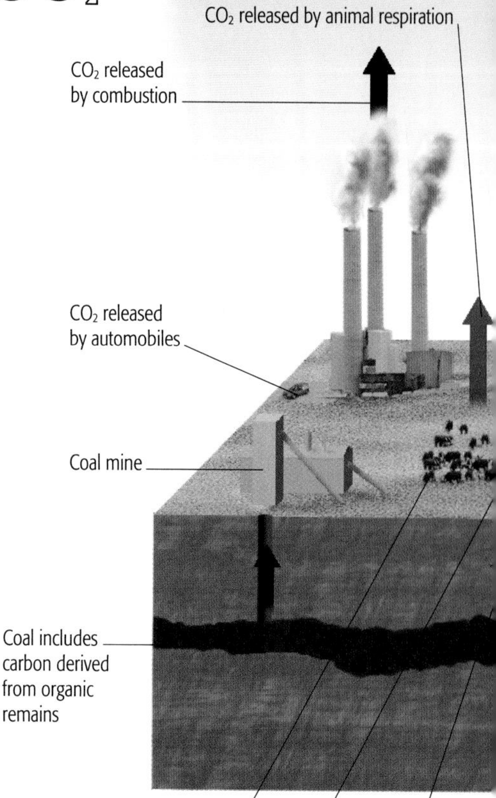

WHERE DID ALL THE CO_2 GO?

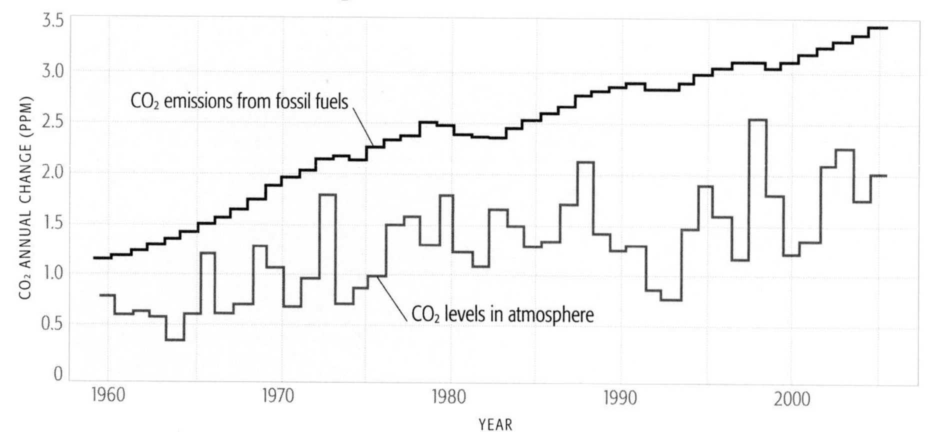

Note that the atmospheric increase does not reflect total emissions; some CO_2 has been has dissolved into the ocean or stripped from the atmosphere during photosynthesis.

THE CARBON CYCLE

The main reservoirs of carbon are the atmosphere, the ocean, and vegetation, soils, and detritus on land. Marine life represents a very small carbon reservoir. On multi-millennial time scales, geologic reservoirs also become important. Various processes transfer carbon between these reservoirs, including photosynthesis and respiration, ocean–atmosphere gas exchange, and ocean mixing. The red arrows show the pathway of fossil-fuel CO_2 from its release to its uptake by vegetation or the ocean.

CO_2 released by plant respiration

CO_2 absorbed by photosynthesis

Carbon stored in plant tissues

CO_2 released by volcanic eruption

CO_2 dissolved in water

Rivers carry eroded carbon to the ocean

Air-sea CO_2 exchange

CO_2 released by combustion

Oil and gas extraction

CO_2 released by marine organism respiration

Carbon in remains of organisms

Sediment

Carbon in sediment turns into limestone and organic-rich shale

Oil and gas

Carbon moves from sediment to oil and gas

Carbon released by decomposition of plants

CO_2 in rain weathers rocks

Carbon buried in plant remains

CO_2 absorbed by photosynthesis

Carbon released by marine organism decomposition

KEY

- Carbon movement
- Photosynthesis
- Weathering and erosion
- Human carbon transformation

Phytoplankton

Marine phytoplankton like these diatoms play an important role in transferring fossil-fuel carbon dioxide from the atmosphere into the ocean.

Positive feedbacks prevail

Despite nature's best efforts to counter our impact on the planet, atmospheric carbon dioxide levels have nonetheless continued to rise, and the planet is warming. Unfortunately, warming reduces nature's ability to absorb carbon dioxide. A number of feedback loops (see p.24) are involved, both positive (enhancing warming) and negative (reducing warming), but positive feedbacks prevail on all but multi-millennial timescales.

Warmer land

Positive feedback: Soil microorganisms increase their growth and respiration rates as their environment warms. One of the waste products of their metabolism is CO_2. As a result, carbon in soils is now being converted to CO_2 at increasing rates.

Negative feedback: This release of CO_2 to the atmosphere by soil microorganisms offsets gains made by plants responding favorably in their growth to elevated CO_2 levels (so-called "CO_2 fertilization"; see p.105).

Warmer ocean

Positive feedback: A warmer ocean has less ability to absorb carbon dioxide, just as an opened can of warm soft drink loses its carbonation and goes flat.

Ocean acidification

Negative feedback: Acidification of the surface ocean (see p.114) reduces the production of calcium carbonate (limestone/ $CaCO_3$) by organisms such as corals and tiny plankton. When these organisms grow their $CaCO_3$ skeletons (i.e., to calcify), CO_2 is released to the water. So calcification reduces the ocean's ability to take up fossil-fuel CO_2. However, it is predicted that some calcifying organisms will become extinct this century. If calcifying plankton and corals become less abundant, then the resulting reduction in the rate of calcification will slightly increase the ocean's ability to take up carbon dioxide.

Goings-on down under
Earthworms, bacteria, and fungi consume plant matter buried in soil, releasing CO_2 that diffuses into the atmosphere above.

Soil microorganisms
Bacteria, such as these shown here, are microorganisms that decompose organic matter in soils.

Pump problems

Positive feedback:
Calcium carbonate is a relatively dense mineral that acts as ballast once an organism dies, carrying its decaying tissue to great depths in the ocean. This "pump" of carbon removes CO_2 from surface waters, allowing more fossil-fuel CO_2 to be absorbed. Loss of the ballast via ocean acidification reduces the ocean's ability to take up atmospheric CO_2.

A sluggish ocean

Positive feedback:
A slowing of ocean circulation in response to global warming reduces the mixing up of nutrients at the ocean's surface, which slows biological productivity. This weakens the action of the biological pump, further reducing the ocean's ability to absorb CO_2.

Rock weathering

Negative feedback:
Increased temperatures and rainfall stimulate the weathering of rocks on land (the process that turns rock into soil and dissolved salts in rivers). Atmospheric carbon dioxide dissolved in rain forms carbonic acid, which aids the rock-weathering process. Increased weathering, therefore, removes CO_2 from the atmosphere.

Conclusion

Models that simulate both the carbon cycle and climate have been run with some, but not all, of these feedbacks taken into consideration. The overall effect of these feedbacks is a more rapid buildup of atmospheric CO_2, and a warmer climate. For example, adding carbon-cycle feedbacks to the A2 fossil-fuel emission scenario (see p.86) leads to an additional 20–220 ppm CO_2 by 2100, and an additional warming of more than 1°C. This additional warming is reflective of a carbon cycle that is approaching the limit of its ability to absorb fossil-fuel emissions.

Melting ice and rising sea level

Sea level is predicted to rise with global warming for two reasons. First, water, like most liquids, expands as it warms. A small rise of 0.1–0.4 m is predicted by 2100, depending on the emissions scenario, due to this effect alone. Second, melting ice is likely to have a major impact on the sea level. It is important to note, however, that not all ice plays an equal role here. The disappearance of high-latitude sea-ice (see p.138), while significant in its own right, will not be a contributor. Much as melting ice cubes in a glass of water do not cause the level of water to rise, melting sea-ice will not cause the sea level to rise. On the other hand, melting continental ice will definitely contribute to a sea level rise.

Significant rise

The continental ice resides in two basic forms. First, there are the permanent ice caps and glaciers in mountain ranges at high latitudes, and even at equatorial latitudes (see p.58). Melting all of this ice, however, would only add at most a sea level rise of about 0.5 m.

More significant are the Greenland and Antarctic continental ice sheets. There is evidence that significant melting of the Greenland ice sheet is already underway, but the rate of future melting is difficult to estimate. Model simulations indicate that local warming over Greenland is likely to exceed 3°C by 2100 in the A1B "middle of the road" scenario (see p.92). Current ice-sheet models indicate that such a warming could lead to the eventual irreversible melting of the Greenland ice sheet, resulting in roughly 5–6 m of global sea level rise. Melting of the most unstable part of the Antarctic ice sheet (the West Antarctic ice sheet) could add an additional 5 m. The models suggest that the completion of this melting could take a number of centuries.

Even faster than we thought

But even state-of-the-art models do not account for some newly observed effects that scientists now believe could significantly accelerate the rate of melting. For example, recently it has been discovered that crevices (called "moulins") are forming in melting continental ice. These moulins allow surface meltwater to penetrate deep into the ice sheet and lubricate the base, allowing large pieces of ice to slide quickly into the ocean. If this phenomenon becomes increasingly widespread, it could lead to a far more rapid disintegration of the ice sheets than predicted by any current models.

The sea level is currently projected to rise between 0.5 m and 1.2 m by 2100.

GREENLAND'S MELTING CONTINENTAL ICE SHEET

This map shows changes in the extent of the region of summer melting in Greenland.

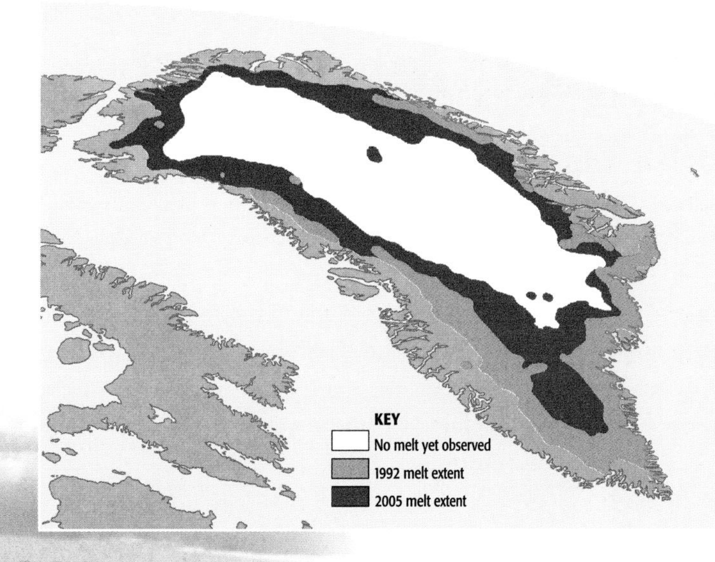

Moulin in Greenland
Crevices (moulins), such as this one in Greenland, allow surface meltwater to penetrate deep into ice sheets.

PROJECTED SEA LEVEL RISE

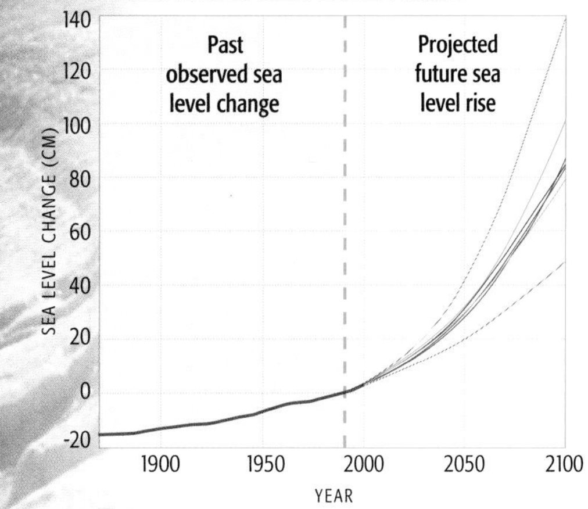

By relying on observations of the past relationships between global temperature and rates of sea level rise, and global temperature projections for the next 100 years (see p.88), scientists can make projections about future sea level rise. The predicted rise by 2100 is between 0.5 m and 1.2 m, depending on the prevailing emissions scenario (see p.86).

Future changes in extreme weather

Frosted roofs
Familiar cold-weather scenes, like these abandoned farmhouses covered with frost near Maupin, Oregon, will become less common as our climate warms.

It is likely that as climate changes, the frequency and intensity of extreme weather events will change. For certain extreme weather events, such as severe frosts and extended heat waves, the science is fairly definitive, and the predicted changes are fairly intuitive. The greater the amount of warming, the more pronounced these trends will be. Changes are predicted to vary regionally.

New trends can be expected to emerge for various types of extreme weather, including heat waves, heavy downpours, and frosts. In the maps below and on the following pages, the colors represent a relative scale. Variations within color fields indicate regions where climate models predict a statistically significant increase or decrease in the quantity in question.

Fewer frosty days

- As temperatures warm, the probability of frosts (nights when temperatures dip below freezing) will decrease markedly.
- The greatest decrease in frost days is likely to occur in regions such as interior North America and Asia, where winter temperatures have been traditionally the coldest.

CHANGES IN FROST DAY FREQUENCY

-1 −0.75 −0.5 −0.25 0 0.25 0.5 0.75 1 1.25 >

This map shows projected changes in the occurrence of frost days by the late 21st century (2080–2099) relative to the observed frequency of occurrence in recent decades (1980–1999). A relative scale is used, where a single unit ("1") represents the typical range of year-to-year variations. The term "frost days" refers to the number of days in the year when the minimum nightly temperature drops below freezing.

More heat waves

- Heat waves (very high temperatures sustained over a number of days) are likely to become more intense, more frequent, and longer-lasting.
- The greatest increase in heat waves is predicted to occur in areas such as the western US, North Africa, and the Middle East, where feedback loops associated with decreased soil moisture may intensify summer warmth (see p.90).

Dry corn
An increase in the frequency of blistering heat waves associated with climate change may make scenes like this one more common in years to come.

CHANGES IN HEAT WAVE FREQUENCY

0 0.75 1.5 2.25 3 3.75 >

This map shows projected changes in the occurrence of heat waves by the late 21st century (2080–2099) relative to the observed frequency of occurrence in recent decades (1980–1999). A relative scale is used, where a single unit ("1") represents the typical range of year-to-year variations. A "heat wave" is a minimum of five consecutive days when the high temperature is at least 5°C above the average.

Wet days and dry days

Most model simulations also indicate that increases are to be expected in the frequency of very intense rainfall events and corresponding flooding. These changes are due to the more vigorous water cycle that will accompany a warmer climate, with greater rates of evaporation from a warmer ocean (see p.24) balanced by more intense precipitation events. Seemingly paradoxical, while many regions are likely to become drier, it is predicted that even in those regions individual rainfall events will become more intense, although longer dry spells will separate them.

CHANGES IN RAINFALL INTENSITY

−1 −0.75 −0.5 −0.25 0 0.25 0.5 0.75 1 1.25 >

This map shows projected changes in the pattern of precipitation intensity by the late 21st century (2080–2099) relative to the observed frequency of occurrence in recent decades (1980–1999). A relative scale is used, where a single unit ("1") represents the typical range of year-to-year variations.

CHANGES IN MAXIMUM DRY SPELL LENGTH

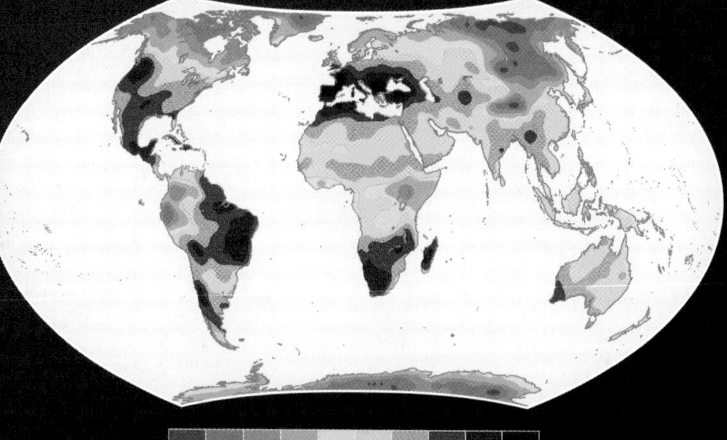

This map shows projected changes in the occurrence of Changes in Maximum Dry Spell Length by the late 21st century (2080–2099) relative to the observed frequency of occurrence in recent decades (1980–1999). A relative scale is used, where a single unit ("1") represents the typical range of year-to-year variations.

Roadway under water
Here a motorist contemplates driving through a flooded road in Beaumont, Texas, following Hurricane Rita on September 24, 2005. Heavy rainfalls will become more common as the atmosphere

Severe storms

It is more difficult to determine how extreme weather events, such as tornados, severe thunderstorms, and hailstorms, will change. This is because such phenomena involve processes that occur at too small a scale to be reproduced in most model simulations. However, it is likely that even if such events do not become more severe or more common in general, individual storms will be associated with more severe downpours and more common flood conditions. This is due to the greater amount of water vapor that a warmer atmosphere can hold. Consequently, this additional water vapor is available to produce rainfall during storm conditions.

Hurricanes and cyclones

But what about the hurricanes and tropical cyclones themselves? We know that there has been a recent trend towards more intense hurricanes in certain basins, such as the tropical Atlantic basin, and that these trends closely mirror warming ocean surface temperatures (see p.56). Also, warmer oceans, all other things being equal, are likely to fuel more intense tropical cyclones, with stronger sustained winds. Model simulations indicate a likely shift towards the strongest (Category 4 and 5) tropical cyclones over the next century, given projected climate changes. There are some important unanswered questions, however. For example, we know that El Niño events change wind patterns over the tropical Atlantic region in such a way as to create unfavorable conditions for tropical cyclone and hurricane formation. And there is still considerable uncertainty about how El Niño will change in response to climate change (see p.90). Climate scientists are currently working to resolve such open questions.

Gusty tempest
Hurricane Dennis' powerful winds, with gusts up to 260 kmh, tossed debris around Cienfuegos, Cuba, on July 8, 2005. Category 4 storms like Dennis (and even stronger Category 5 storms) may become more frequent as tropical sea surface temperatures warm with climate change.

Stabilizing atmospheric CO_2
Is a greenhouse world a better world?

The rapid rise of atmospheric carbon dioxide levels over the last two centuries is a clear indication of society's hunger for cheap energy, a hunger that is insatiable and which is being fed an unhealthy diet of fossil fuels. Recognizing this, scientists have been studying future fossil-fuel-use scenarios with substantially elevated atmospheric CO_2 levels. The IPCC Fourth Assessment Report paid specific attention to scenarios that had stabilization targets from one-and-a-half to three-and-a-half times the pre-industrial value of 280 ppm; these values are considerably higher than the 386 ppm average of 2008.

The need to act now

A broad range of CO_2 stabilization targets has been studied, ranging from 450 ppm to 1000 ppm. The bottom graph at right shows the gradual climb to these levels over the next three centuries. Some interesting characteristics of these curves emerge:

- The lower the stabilization target, the sooner peak emissions of fossil fuel carbon dioxide must occur. In other words, the lower the level at which we want to stabilize the CO_2 levels in the air, the sooner we have to cut back on fossil-fuel use. To stabilize atmospheric CO_2 levels at 450 ppm, we would need to reach peak usage before 2020.
- Lower stabilization levels can be achieved only with lower peak emissions; while the 1000 ppm target allows CO_2 emission rates to double, the 450 ppm target allows them to increase by only 50% or so.
- All stabilization targets require sharp reductions in CO_2 emissions following the peak. Low stabilization targets require that emission rates fall below the current rate within a few decades.

EMISSIONS SCENARIOS FOR CO_2 STABILIZATION

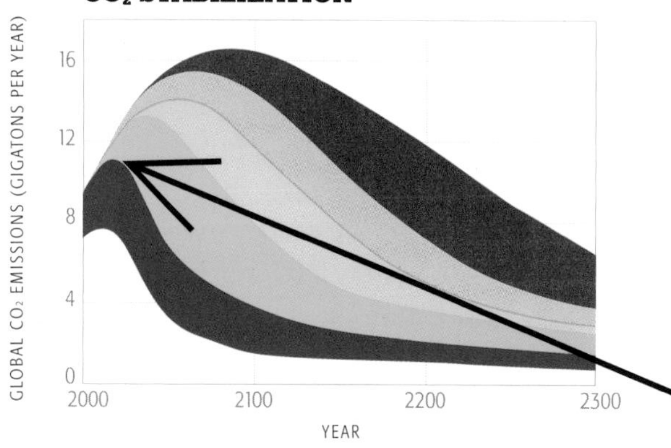

The level at which emissions peak determines the level at which atmospheric CO_2 stabilizes.

CO_2 STABILIZATION TARGETS

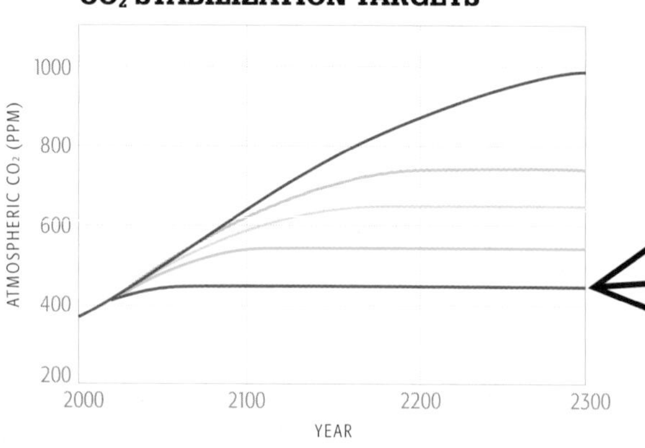

The emission scenarios on this graph show different possible stabilization levels of atmospheric CO_2 over the next three centuries. The colors of the lines in this graph correspond to the colors that represent the emissions scenarios in the graph above.

It is also interesting to note that the projected climate changes associated with even the most conservation-minded emission targets are substantial. The warming projected for 450 ppm is 2.1°C, an increase that could lead to a sea level rise of half a meter or more by the end of the 21st century (see p.98). Moreover, even with the CO_2 level stabilized, the temperature and sea level will continue to rise as the sluggish climate system adjusts to the new atmospheric composition, committing us to continued warming and coastal inundation for decades to come.

Will more CO_2 benefit plants?

Given the serious reductions that would be required to achieve low stabilization targets, some people argue that the higher CO_2 stabilization targets shouldn't be considered failures but rather desirable objectives for beneficial climate modification. This line of reasoning argues that plants require CO_2 for photosynthesis and that more CO_2 should benefit plants (CO_2 fertilization), including the crops that feed the people of the world. Crops grown under ideal conditions in greenhouses with elevated carbon dioxide levels do outperform those grown under ambient atmospheric conditions. In the presence of elevated CO_2, plants do not have to open their pores as wide; this reduces water loss and infection by germs, and encourages growth. However, these benefits cannot be fully realized in nature if other factors, such as a lack of nutrients or inadequate soil, are limiting growth.

The existence of growth-limiting factors, combined with other negative impacts from elevated CO_2 levels—such as ocean acidification and loss of coral reefs (see p.114)—suggests that the "greening of planet Earth" may not be an achievable or desirable outcome of fossil-fuel burning.

To stabilize atmospheric CO_2 levels at 450 ppm, fossil fuel use needs to peak by 2020

With atmospheric CO_2 levels at 450 ppm, global temperature increases by 2.1°C and sea level rises by up to half a meter or more

Part 3
The Impacts of Climate Change

Recent studies indicate that the future impacts of climate change are likely to be far more significant than those observed to date. Human societies, natural habitats, and a myriad of animal and plant species will all be affected by changes in temperature and precipitation patterns in the decades ahead. The precise impacts will depend on the rate and amount of warming, and on the adaptive measures taken by society.

The rising impact of global warming

A world under stress

A list of the potential impacts of global warming on humanity and planet Earth makes for sobering reading. These impacts include a greater tendency toward drought in some regions, the widespread extinction of animal species, decreases in global food production, the loss of coastline and coastal wetlands, increased storm damage and flooding in many areas, and a wider spread of infectious disease. Stresses such as these could, in turn, lead to increased competition for natural resources, overtaxed social services and infrastructures, and conflict between regions and nations. Sustainable approaches to development will be necessary to decrease the vulnerability of society, ecosystems, and the environment to future changes.

Dry lake bed
Scenes like this one from Death Valley, California, may become ubiquitous in the southwestern US if persistent widespread drought takes hold in a warmer world.

GLOBAL WARMING IMPACT SCALE

EFFECT OF FURTHER TEMPERATURE CHANGES

2010 2020 2030 2040 2050 2060 2070 2080 2090 2100

+5°C ← Global economic losses of up to 5% of GDP

- At least partial melting of Greenland and West Antarctic ice sheets, resulting in eventual sea-level rises of 5–11 m

+4°C ← Substantial burden on health services

- Decreases in global food production
- About 30% of global coastal wetlands lost
- 40% - 70% of species extinct
- Corals extinct

+3°C ← Changes in natural systems cause predominantly negative consequences for biodiversity, water, and food supplies

- Millions more flood victims every year
- Major loss of tropical rainforests

+2°C ← Human mortality increases as a result of heat waves, floods, and droughts

← 9% - 31% of species extinct

- Widespread extinction of amphibians underway

+1°C ← Decreases in water availability; more frequent droughts in many regions

- Wildfire risk increases, as do flood and storm damage
- The burden from increased incidence of malnutrition and diarrhoeal, cardio-respiratory, and infectious diseases escalates

+0°C ← **Amount of global warning** (°C increase over 1980-1999 levels)

Is it time to sell that beach house?

Ten percent of the world's population lives in coastal and low-lying regions, where the elevation is within 10 m of sea level. In some places, such as Bangladesh, this population figure is nearer to 50%. Rising sea level (see p.98), increasing destruction associated with tropical storms (see p.56), increasing coastal erosion, and larger wave heights all pose serious threats to coastal and low-lying regions.

Soggy cities

The most obvious threat associated with global sea level rise is coastal inundation. Coastal regions across the globe are potentially at risk. In North America, for example, significant loss of land on the mid-Atlantic and northeast coastlines could occur with just 6 m of sea level rise—an amount of sea level rise that would result from the melting of the Greenland ice sheet (see p.98). Substantial portions of Europe's "low countries" of Belgium and the Netherlands would also be submerged with a sea level rise of 4–8 m.

SOUTHERN FLORIDA

Southern Florida would be submerged if it were to experience between 4 m and 8 m of sea level rise.

1-meter rise

2-meter rise

4-meter rise

KEY
Land submerged if sea level rises

MISSOURI
KENTUCKY
VIRGINIA
TENNESSEE
NORTH CAROLINA
ARKANSAS
SOUTH CAROLINA
Georgetown
Charleston
Savannah
LOUISIANA
ALABAMA
GEORGIA
MISSISSIPPI
TEXAS
New Orleans
Port Arthur
Galveston
Morgan City
Tampa
FLORIDA
GULF OF MEXICO
Naples
Miami
8-meter rise

NORTHEAST COASTLINE

Most of New York City and Boston could be submerged if sea level were to rise by 6 m.

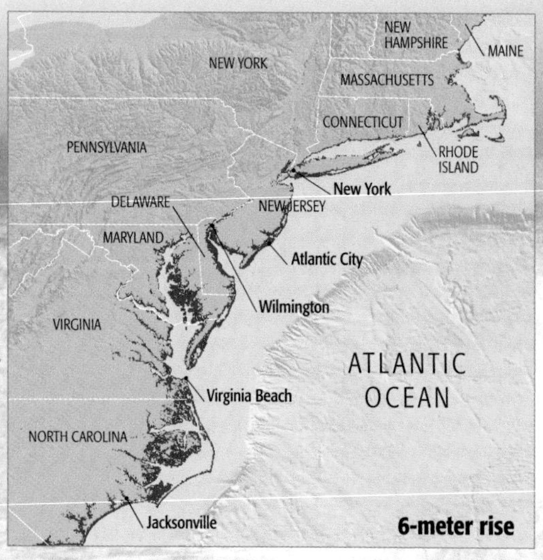

Human loss

Even those coastal regions not inundated by higher sea levels will be subject to increased exposure to flood and storm damage, more intense coastal surges, and altered patterns of coastal erosion. Associated impacts are likely to include loss of human life, damage to human infrastructure and real estate, degraded water quality, and decreased availability of fresh water due to saltwater intrusion. Coastal habitats will be lost if water levels and wave heights substantially increase. Significant population displacement will also be a factor. Communities, habitats, and economies on all of the major continents will be affected by even just 1 m of sea level rise. The cost rises dramatically at 5 m and 10 m. The melting of the Greenland ice sheet will likely lead to 5–6 m of sea level rise, and the melting of a large part of the West Antarctic ice sheet would probably result in at least an extra 5 m (see p.98).

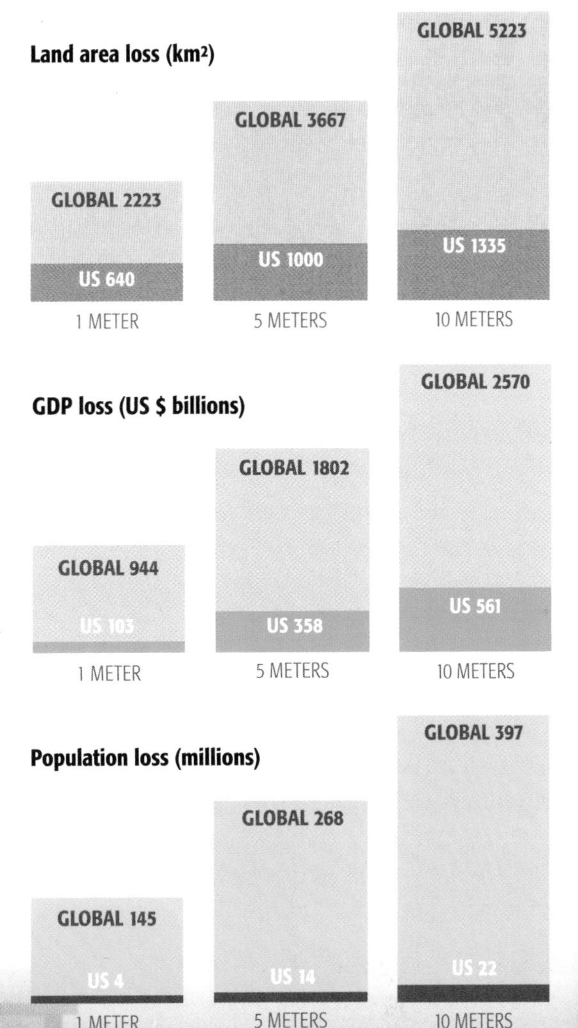

GLOBAL LOSSES

These bar graphs show the impacts of sea level rises of 1 m, 5 m, and 10 m. Total global and US losses are shown. Note that losses are sizeable even in the event of 1 m of sea level rise, which could plausibly occur by the end of this century.

Rough tides

The waves generated by Hurricane Felix encroached on evacuated beach houses on Virginia Beach, Virginia, in August 1995. Homeowners in this community face the dual threat of stronger storms and rising sea level as a result of climate change.

Ecosystems Worth saving?

Perhaps your drive to work takes you through a reeking, swampy area that's often subject to flooding. Imagine a future in which this lowland quagmire is filled and a new interstate is built, conveying you rapidly through the area with your windows closed and your eyes fixed straight ahead. An improvement? In some ways, perhaps. But was value lost when the ecosystem was destroyed? Is the attractive pond the transportation department built when it reshaped the area serving the same function as the wetland that was destroyed? This begs a further question: are wetlands and other natural ecosystems of any real use to us?

Why we need wetlands

Wetlands provide an important service to their surroundings. Storm waters flowing into a wetland lose energy and spread out across a broad area, thus reducing flooding downstream. Furthermore, sediment and contaminants, such as iron and acidity from mine runoff and nitrogen from farm fertilizers, are removed as water percolates through wetlands before entering our drinking and irrigation water. Wetlands are also biologically diverse ecosystems, providing homes for endangered species and a refuge for migrating birds. This makes them ideal places for hiking, bird-watching, canoeing, and fishing.

What is an ecosystem?

An ecosystem consists of interdependent communities of plants, animals, and microscopic organisms, and their physical environment. All these different elements interact and form a complex whole, with properties that are unique to that particular combination of living and non-living elements. Ecosystem boundaries are generally delineated by climate: desert ecosystems in the subtropics, tropical rainforest ecosystems near the equator, and tundra ecosystems near the poles. As climates have changed in the geologic past, ecosystems have shifted in response. But past climate changes were slower than the projected future changes. Will the ecosystems of today be able to adjust their boundaries as the climate changes, or will they be stranded with incompatible climates? And why, for that matter, should it concern us?

Worth saving

Ecosystems are valuable to humanity because they assist us with:

- **Provisions:** food (seeds, fruits, game, spices); fiber (wood, textiles); medicinal and cosmetic products (dyes, scents)
- **Environmental regulation:** climate and water regulation; water and air purification; carbon sequestration; protection from natural disasters, disease, and pests
- **Cultural benefits:** appreciation of, and communion with, the natural world; recreational activities

Ecosystems are reasonably resilient to change, including a modest amount of human disturbance. But there are limits to this resilience, and human activity is already pushing those limits in some ecosystems (see p.114). Climate change not only challenges the persistence of ecosystems: it may also lead to the extinction of species that cannot adapt or migrate sufficiently rapidly (see p.118).

Unlike local highway construction, the stress placed on ecosystems by climate change is an insidious one—less obvious, but perhaps more permanent. As climate regimes shift poleward, ecosystems will likely follow. However, they may be stopped from doing so by natural factors (such as incompatible soils) and human development (roads, cities, agriculture). The result could be widespread destruction of ecosystems, with attendant loss of benefits to society and a significant reduction in global biodiversity.

Humans meet nature
"Alligator Alley", the highway through the Florida Everglades wetlands ecosystem, was designed with numerous underpasses to minimize environmental impact and threats to the surrounding ecosystems.

Coral reefs Will ocean acidification be their demise?

Coral reefs are among the world's most diverse ecosystems. On a single snorkeling adventure in a healthy reef you can see more species of animals than you can during a lifetime of hiking in mid-latitude forests. Reefs also provide food for hundreds of millions of people, a barrier of defense against the ravages of hurricanes and tsunamis, and a tremendous source of tourism income for nations lucky enough to have them grace their coastlines.

But all is not well with coral reefs, and scenes like the ones depicted on the following pages are becoming all too common. Studies conducted on reefs throughout the world document widespread reductions in coral coverage. The National Oceanic and Atmospheric Administration (NOAA) estimates that 10% of coral reefs are already damaged beyond recovery, and that 30% are in critical condition and may die within the next 10 to 20 years.

Healthy reefs
Coral reefs are among the most diverse and productive ecosystems on Earth. Unfortunately, healthy reefs such as the one shown here are becoming increasingly rare.

Unless significant measures are taken to reduce the stress on coral reefs from human activities, 60% of the world's coral reefs may die by the year 2050.

Unhealthy reefs
This coral reef is bleached: the corals have lost their symbiotic algae that give them their characteristic color. Coral bleaching is believed to be caused by excessively hot ocean temperatures.

EFFECT OF DISEASE ON CORAL HEAD

1988

1998

Black band disease, a bacterial infection linked to warmer water temperatures, destroyed this massive coral head in the Florida Keys. The bacteria are concentrated in the black band, which migrates outward over time, leaving dead coral behind.

Causes of decline

The causes of coral reef decline are many, and include:

Natural stressors

- Disease
- Predation
- Out-competition by algae

Human activities

- Overfishing
- Pollutant runoff from land
- Careless snorkelers walking on delicate coral
- Fuel leaks
- Fuel and wastewater discharge from boats, and oil spills

Additional natural factors are being exacerbated by human activity. For example, coral bleaching—the loss of the algae that live in a symbiotic relationship with the coral animal and give it its color—has been directly linked to intervals of exceptionally hot ocean temperatures. Human-induced global warming is likely contributing to this problem.

Marine protected areas

Marine protected areas (MPAs) are being established around the world, and have proven to be effective at staving off coral and fish losses. They have been shown to be of great economic benefit as well. The United Nations Environmental Program estimates that MPAs cost less than US $1000 per square kilometer, whereas the economic value of coral reefs has been estimated at US $100,000–$600,000 per square kilometer.

Unfortunately, MPAs cannot protect coral reefs from warming, nor can they protect corals from the direct effect of increases in atmospheric carbon dioxide. The ocean has absorbed approximately half of the carbon dioxide released by fossil-fuel burning and deforestation (see p.94). When carbon dioxide dissolves in water, it is transformed into carbonic acid, which makes the water less conducive to coral growth. Sadly, a recent study has shown that every square kilometer of the ocean, with the possible exception of a few remote Arctic areas,

Experimental studies suggest that if fossil-fuel burning rates continue to increase, corals will be unable to grow skeletons by the end of this century.

has suffered detrimental consequences from human activities similar to those experienced by coral reefs.

The only way to prevent this catastrophe is to reduce or eliminate CO_2 emissions, or to develop effective means of sequestering CO_2 before it escapes into the atmosphere.

Coral or cars?

We face a stark choice: either we continue emitting carbon dioxide at ever-increasing rates and accept the demise of coral reefs, or save the coral reefs by reducing or eliminating fossil-fuel CO_2 emissions. It's a genuine dilemma, but for those who value the beauty and understand the importance of coral reefs to the global environment and to society, the path is clear.

And it's not just coral reefs that are at stake. A common theme of this book is

CORAL GROWTH RATE IN DECLINE

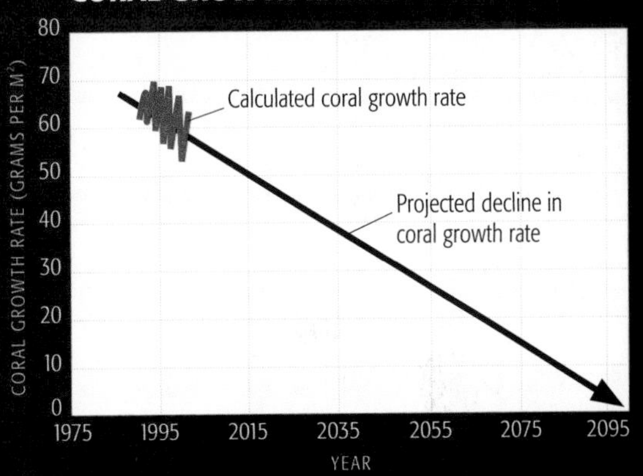

This graph shows the projected decline in coral skeleton growth rate (calcification) through this century. The data shown in red represent a calculation of coral growth rate (in grams of skeleton produced per square meter of coral per year) based on observed changes in CO_2 buildup in the ocean near Hawaii. The black line shows the extrapolated trend for an A1B "middle of the road" scenario (see p.86). If greenhouse gas emissions continue to escalate, corals won't be able to grow by the year 2100 or so.

The highway to extinction?

The diversity of species on planet Earth today is the result of millions of years of evolutionary interaction between life and its environment. Human intervention is a new, powerful force, which some liken to the forces that led to mass extinctions of life in the past, such as the asteroid impact that probably precipitated the demise of the dinosaurs 65 million years ago.

Polar bears in danger

A case in point is the precarious future of the polar bear, which depends on expansive sea-ice cover to reach and feed on seals. The earlier spring break-up and retreat of the sea-ice is now forcing polar bears to remain on the tundra, where they must fast and survive on reserves of fat. This puts particular stress on female polar bears, which spend the winter in nursing dens and need easy access to seals in spring to rebuild their fat reserves.

Temperature changes and limited water availability can stress individual organisms, which find themselves suddenly outside of their climate "comfort zone." Typically it is not just one species that is affected. A new report indicates that warmer temperatures are translating into lower fish populations off Antarctica. This, in turn, is resulting in lower survival rates for king penguins, threatening a potential population collapse over coming decades. If global warming means that penguins are not getting enough food, then the conditions for the organisms below them in the food chain are probably even worse.

Extinct amphibians

Also of concern is the worldwide loss of amphibians. In a recent assessment, 122 species of amphibians were listed as "possibly extinct" and another 305 as "critically endangered." The golden toad was last seen in the cloud forests of Costa Rica in 1988. In cloud forest ecosystems, mist from clouds is the primary source of moisture. As the climate warms, the trade winds rising up the mountain slope condense at higher elevations, so the clouds shift upwards. This leaves the forests less cloudy and drier, with warmer nights. Birds, reptiles, and amphibians have all been affected, but the golden toad and the harlequin frog are now believed to be extinct. One theory posits that warmer nights favor the growth of the chytid fungus—a potentially fatal pathogen that grows on amphibian skin.

Adapt or die

Amphibians are the first group of organisms identified as at risk of extinction from global warming. Many more will follow as the planet warms, especially if the rate of warming is rapid. Organisms adapt and ecosystems migrate at rates that sadly may be too slow to prevent ecosystem collapse and the extinction of species.

Bears on ice
Since polar bears hunt on sea-ice, the melting of the Arctic ice cap is making it increasingly difficult for the bears to find sufficient food.

Gone for good?
The golden toad was last seen in the cloud forests of Costa Rica in 1988.

The IPCC report states with medium confidence that 20–30% of plants and animals will be subject to increased risk of extinction if global temperatures rise to 2°C above the pre-industrial level and perhaps 40–70% of species will be at risk of extinction if temperatures rise by 4°C.

We must remember that extinction is irreversible, and that we are inseparably dependent on the diversity of species harbored by our planet, and the goods and services provided by the ecosystems they support (see p.112).

BIODIVERSITY IMPACT SCALE

EFFECT OF FURTHER TEMPERATURE CHANGES

+4°C ← **40%–70% of species extinct**
- Corals extinct
- Ecosystems lose 7-74% of areal extent

+2.9°C ← **21%–52% of species extinct**
- Major loss of tropical rainforests, with biodiversity losses from climate change exceeding those due to deforestation

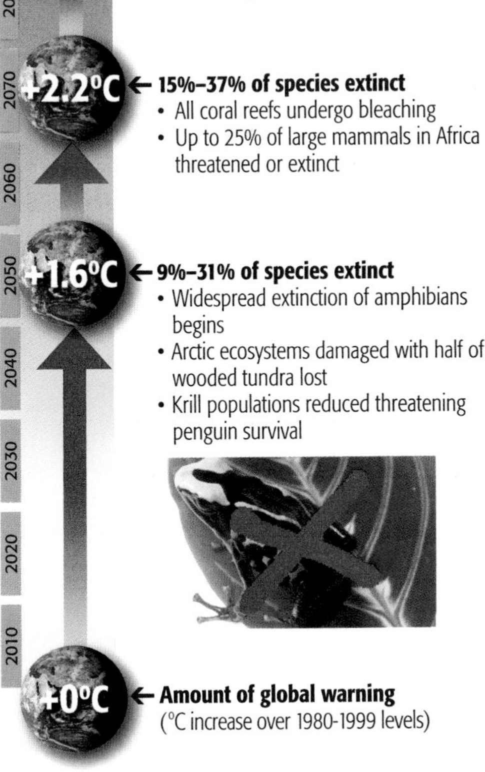

+2.2°C ← **15%–37% of species extinct**
- All coral reefs undergo bleaching
- Up to 25% of large mammals in Africa threatened or extinct

+1.6°C ← **9%–31% of species extinct**
- Widespread extinction of amphibians begins
- Arctic ecosystems damaged with half of wooded tundra lost
- Krill populations reduced threatening penguin survival

+0°C ← **Amount of global warning**
(°C increase over 1980-1999 levels)

Profile James Lovelock

"... for now, the evidence coming in from the watchers around the world brings news of an imminent shift in our climate towards one that could easily be described as Hell: so hot, so deadly that only a handful of the teeming billions now alive will survive."

This most dire of predictions comes from James Lovelock, a British inventor and scientist, in his book *The Revenge of Gaia* (Penguin, 2006). Lovelock is perhaps best known for his "Gaia Hypothesis." According to this hypothesis, all life on Earth, acting like a single organism, participates in planetary-scale regulation of the climate, atmosphere, and oceans. Although controversial, Lovelock has displayed an uncanny ability to originate ideas that evolve from fringe to mainstream.

Working on the NASA Viking mission in the 1960's, Lovelock was responsible for helping NASA determine if life exists on Mars. Believing his colleagues were too

"Earth-centric" in their attempts to determine the existence of extraterrestrial life, Lovelock instead pondered the more general question of what the common characteristics of life might be, and how those characteristics might be expressed in detectable ways on another planet. On Earth, for example, plants produce oxygen and bacteria produce methane (natural gas) in such large amounts that Earth's atmosphere has considerable quantities of both gases, much more than would co-exist on a lifeless planet. It follows that the presence of methane and oxygen in Earth's atmosphere is a signature of a living planet. Lovelock knew that Mars, on the other hand, had a nitrogen–carbon dioxide atmosphere, which is essentially what one would predict of a lifeless planet. Based on this observation, Lovelock asserted that life would not be found on Mars. NASA returned from Mars with important information and fascinating images of our nearest planetary neighbor, but they detected no signs of life. Lovelock's method of planetary life-detection using atmospheric composition is now the central approach of the Terrestrial Planet Finder program at NASA.

Lovelock also conceived of an approach to "terraforming" Mars that would involve melting the ice-caps, which are rich in frozen CO_2, with nuclear warheads. This would release greenhouse gas into the atmosphere and warm the planet to habitable conditions. Unable to publish this idea in scientific journals, Lovelock resorted to writing (with Michael Allaby) about terraforming in an entertaining science-fiction book, *The Greening of Mars* (Warner Books, 1984). Much later, a NASA-sponsored terraforming study arrived at very similar approaches for making the planet habitable.

The list of Lovelock's prescient ideas extends beyond the scope of this book, as does the societal and scientific impact of his many inventions. For example, he suggested that evergreen trees warm the climates of high latitudes by shedding snow and absorbing more sunlight than snow-covered tundra. This feedback loop is now accepted as a standard component of climate models.

We might do well to carefully consider Lovelock's "outrageous" predictions for the future of humanity in a greenhouse world. In some respects, the dire predictions of the Fourth Assessment Report of the IPCC are more in step with Lovelock's views than with the mainstream thought that prevailed within the scientific community just a few years ago.

Too much and too little

Will floods and droughts really get worse?

There is nothing more precious to living things than water. Changes in the availability of fresh water are of paramount importance in gauging the impacts on society of climate change. These impacts may at first seem contradictory, since increased drought is predicted in many regions, while more frequent intense precipitation events and flooding are predicted for others (see p.88). Such diverse changes result from a complex pattern of shifting rain belts, more vigorous cycling of water in a warmer atmosphere, and increasing evaporation from the surface due to warmer temperatures.

Sunken city
The Mississippi River flooded the city of Kaskaskia, Illinois, in summer 1993. The city, already an island in 1993 thanks to past repeated floodings, never recovered from this episode. Flooding is likely to become more commonplace in many parts of North America and Europe, even in places that are simultaneously suffering from drought.

One of the most significant potential impacts of climate change is diminished or unreliable fresh water supplies.

Globally, water demand is likely to escalate significantly in future decades, primarily due to population growth. Yet this growth is taking place at a time when, in many regions, fresh water resources may be growing more scarce due to climate change.

A combination of warmer water, more intense rainfall events, and longer periods of low river and stream flows will also exacerbate water pollution. Combined with other aggravating factors, such as population growth and increased urbanization, these impacts put intense pressure on fresh water supplies. Since steady running water is required for hydroelectric energy plants, and for cooling towers used in nuclear energy production, decreased river and stream flows can threaten energy resources too.

Slow stream
As a result of severe drought in the Brazilian Amazon, rivers that are the lifeline for local people for transport, supplies and trading are drying up, threatening many communities.

WORLDWIDE

For the more than 15% of the world's population that depends on the seasonal melt of high elevation snow and ice for fresh water, the melting of glaciers and ice caps (see p.98) represents a serious threat.

Serious negative impacts

On balance, the negative impacts of changing precipitation patterns outweigh the benefits. For example, the increases in annual rainfall and runoff in some regions are offset by the negative impacts of increased precipitation variability, including diminished water supply, decreased water quality, and greater flood risks. There is hope, however, that in some cases adaptations (e.g., the expansion of reservoirs) may offset some of the negative impacts of shifting patterns of water availability (see p.150).

US

A steady increase in the population of cities such as Phoenix and Las Vegas is occurring at precisely the same time that drought conditions are worsening. In the Pacific Northwest, streamflow may have decreased so much by 2020 that the 2007 level of water demand will not be able to be met, and salmon habitat will be lost.

FUTURE CLIMATE CHANGE IMPACTS ON WATER

SOUTH AMERICA

Aquifers will be depleted 75% by 2050.

AFRICA

The spread of disease will increase due to more heavy precipitation events in areas with poor water supplies and an overtaxed sanitation infrastructure.

SOUTHERN EUROPE AND THE MEDITERRANEAN

Many arid and semi-arid regions, such as the Mediterranean and parts of southern Europe, southern Africa, and much of Australia, are likely to suffer from increased drought. Electricity production potential at hydropower stations may decrease by more than 25% by 2070.

MORE FREQUENT EXTREME DROUGHT EVENTS

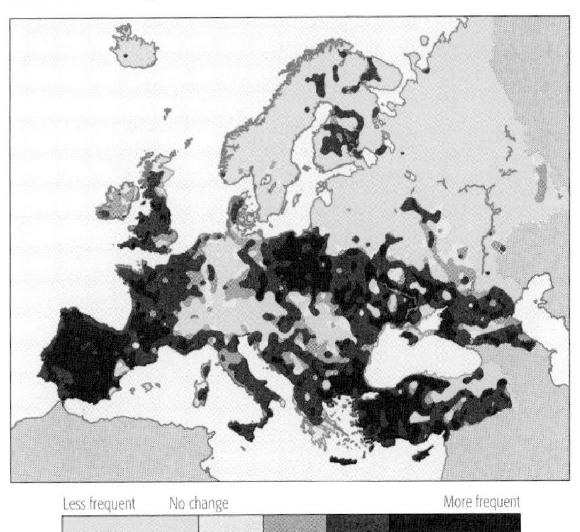

SIMULATED RETURN PERIOD (TYPICAL NUMBER OF YEARS BETWEEN CONSECUTIVE DROUGHTS) FOR EXTREME DROUGHT (I.E., DROUGHT WITH A MAGNITUDE EQUAL TO WHAT IS CURRENTLY CONSIDERED A 100-YEAR DROUGHT) BY LATE 21ST CENTURY (2070–2079).

Climate model simulations predict that the spacing between consecutive extreme drought events (defined as once-in-a-hundred-years events) will decrease sharply by the 2070s for the "middle of the road" emissions scenario (see p.86).

BANGLADESH

Areas with increased rainfall and runoff will suffer from an enhanced risk of flooding. The impacts are likely to be especially harsh for regions like Bangladesh, which is already facing the pressures of rising sea level (see p.98).

INDIA

In many coastal regions there will be plenty of water, but it will be the wrong kind! A sea level rise of 0.1 m by 2040–2080 will threaten the fresh water supply.

< -50 -30 -20 -10 -5 0 5 10 20 30 50 >

MEAN CHANGE OF ANNUAL RUNOFF, IN PERCENT, BETWEEN THE PRESENT (1981–2000) AND 2081–2100 (SIMULATED)

Water resources will shift, but not in society's favor.

Is warming from carbon dioxide leading to more air pollution?

Carbon dioxide isn't a pollutant in the typical sense—that is, something introduced into the environment that is a direct threat to human health or to nature. The most notable detrimental health effects of rising CO_2 levels are indirect. They include the negative health effects related to warming we will discuss in more detail later (see p.132), and the intensification of air pollution—a newly discovered phenomenon.

The latest discovery about air pollution utilizes a comprehensive model of climate, pollutant chemistry, and human health effects to calculate the relationships

buildup of pollutants. By directly calculating human health effects, this model is different from other climate models.

Smog is produced when emissions from incomplete fossil-fuel combustion react to produce pollutants. One pollutant of note is ground-level ozone, a lung irritant that also damages crops, buildings, and forests. (Ozone in the upper atmosphere protects us from ultraviolet radiation, but near the ground it acts as a pollutant.) Warming accelerates ozone production and promotes air stagnation, leading to higher ground-level ozone levels.

Los Angeles sunset
Although significant progress has been made to reduce smog in Los Angeles, greenhouse warming may jeopardize future efforts to clean up the air here and in other large polluted cities, such as Houston and Mexico City.

The comprehensive model indicates that for each 1°C increase in temperature there will be an additional 20,000 pollution-related deaths worldwide. This same amount of warming results in even more notable increases in the incidence of asthma and other respiratory illnesses. In the model simulation, there was no question that the cause of these health problems was the buildup of CO_2, because that was the only change to which the model was subjected. If these predictions are correct, society must reckon with yet another unanticipated consequence of fossil-fuel burning: smoggier, unhealthier cities.

In the US, every 1°C temperature rise will result in 1000 extra pollution-related deaths.

War

Some climate change skeptics believe that global warming is the sole domain of conservationists, peaceniks, and utopian idealists. Yet policy experts in national security and global conflict, including former CIA directors and White House chiefs of staff, are also worrying about the potential threats posed by fossil-fuel burning.

Why? First, there is the most immediate and obvious reason: reliance on fossil fuels threatens the security of many developed nations by placing them at the mercy of volatile regimes. And there is another, less-discussed security reason: an open Arctic Ocean, forecast to be the norm in "middle of the road" emissions scenarios (see p.138), would have clear international implications. If the forecasts are correct, North American and Eurasian nations will suddenly have new northern coastlines to defend.

But perhaps the most important security concern is the potential for increased competition among nations for diminishing essential resources. As any student of history can tell you, in many parts of the world, such as Latin America, increased stress on resources has historically led to local unrest and unstable regimes.

The possibilities for conflict are countless. For example, the predicted changes in precipitation patterns will naturally create increased competition for available fresh water. Imagine the current Middle East political strife with the added facet of vicious water-resource competition.

Ice floes and flag

This pairing in Glacier Bay National Park, Alaska, eerily foreshadows the emerging connection between melting ice and the threat to US national security.

Environmental refugees

A future with expanded patterns of drought, and conditions unfavorable for agriculture and farming is likely to change people's notions of what constitutes desirable land. Sea level rise and other factors will also potentially make currently inhabited regions inhospitable to humans, thereby increasing competition for habitable land.

The term "environmental refugee" has already been coined to describe individuals fleeing their homelands for more benevolent conditions and climes. Lest one think this merely a hypothetical concept, it should be noted that an estimated 25 million environmental refugees have already been displaced. This is more people than have fled civil war or religious persecution in recent years. Climate change appears to be driving the ongoing migration from the dry Sahel to

OPENING THE NORTHWEST PASSAGE

As the Artic sea-ice retreats and the once-fabled "Northwest Passage" opens up, new sea routes connecting the north Pacific and north Atlantic Oceans are becoming available.

neighboring regions of West Africa. It also appears to be playing a role in the exodus of people from parts of India, China, Central America, and South Africa.

Ripe for conflict

An optimist might hope that the global threat of climate change will unite the international community as never before, spurring a coordinated campaign among nations to save humanity. Unfortunately, conflict experts foresee the possibility of a different scenario.

A global population predicted to increase to about 9 billion by the mid-21st century, combined with stresses on water, land, and food resources could create the "perfect storm."

As nations around the world exceed their capacity to adapt to climate change, violence and societal destabilization could ensue, leading to unprecedented levels of conflict both between and within nations.

In the "middle of the road" climate change scenario, a combination of worsened drought, oppressively hot tropical temperatures, and rising sea levels could displace a large enough number of people by the mid-21st century to challenge the ability of surrounding nations to accept them, with political and economic turmoil ensuing.

One possible scenario, for example, is that increasingly severe drought in West Africa will generate a mass migration from the highly populous interior of Nigeria to its coastal mega-city, Lagos. Already threatened by rising sea levels, Lagos will be unable to accommodate this massive influx of people. Squabbling over the dwindling oil reserves in the Niger River Delta and the associated potential for corruption will add to the factors contributing to massive social unrest.

Another possible scenario is that drought and decreased river runoff in southwestern North America will strain already water-poor and resource-starved Mexico, leading to increased migration to the US and stress on already delicate diplomatic relations between the two countries.

Even more ominous conflicts can be imagined for the more extreme climate change scenarios. As the nations and peoples of the world compete for diminishing resources, it may become increasingly difficult to establish or maintain stable governance. Indeed, some experts have described worst-case scenarios not so different than those in post-apocalyptic fables such as *Mad Max* and *The Road*.

Famine

More people, less water, less food

Climate change has the potential to seriously undermine the world's food supplies. Sadly, many of the regions most likely to be affected are already finding it difficult to meet existing food demands.

Short-term positives, long-term negatives

Perhaps surprisingly, some regions, such as the US, Canada, and large parts of Europe, stand to benefit at moderate levels of additional warming (1–3°C), thanks to increased crop and livestock productivity.

For these lucky countries, warming will result in longer growing seasons, favorable shifts in rainfall patterns, and higher CO_2 levels, which provide a short-term benefit for plant growth. Indeed, global food production is projected to increase on average with moderate levels (1–2°C) of future warming. But at even those moderate levels, many tropical and subtropical regions—including India and sub-Saharan and tropical Africa, which are already struggling to meet food and pasturing demands—will likely experience a combination of warmer temperatures and decreased rainfall. This will cause a corresponding decrease in productivity of key crops, such as the major cereals.

PROJECTED CLIMATE CHANGE IMPACTS ON CROP AND LIVESTOCK YIELDS

This map illustrates expected changes in crop and livestock yields by 2050 under the "middle of the road" emissions scenario (see p.86). Note that tropical regions suffer more losses as a result of climate change than do temperate regions.

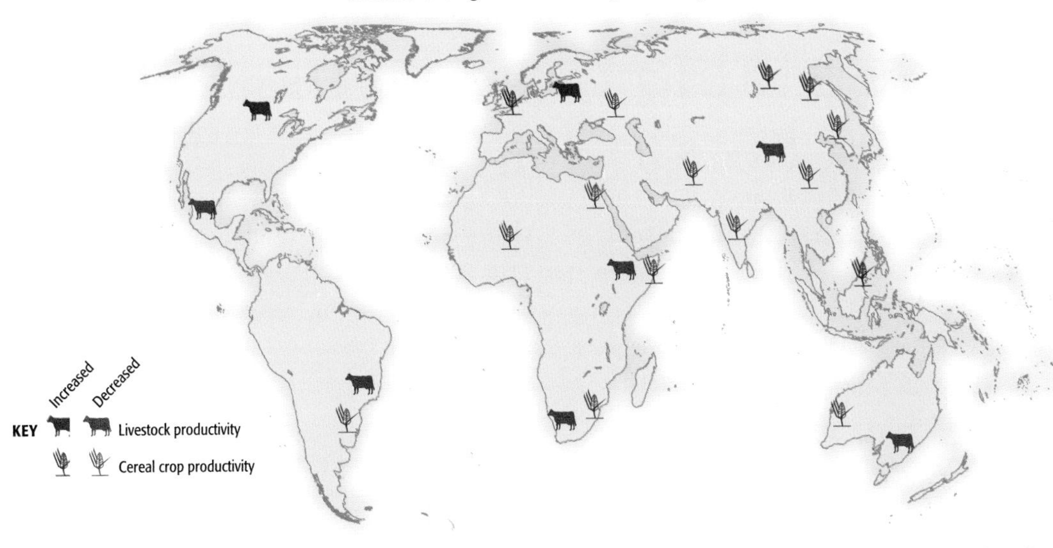

Moreover, the substantial warming (more than 3°C) predicted to beset us as early as the late 21st century in the "middle of the road" emissions scenario (see p.86) will probably have negative impacts for food crops in all major agricultural regions. Increased forest fires (see p.135) and more frequent disease and pest outbreaks (see p.132) could also diminish available food resources. What's more, the agricultural labor supply is likely to be disrupted by population displacement due to flooding (see p.98) and epidemics.

At sea

Fish populations, and thus commercial and subsistence fishing, may be greatly affected by climate change. The impact of warming is likely to be variable, leading to either increases or decreases in aquatic populations, depending on the location and the species of fish present. Changing ocean circulation patterns also represent a wild card. In particular, the potential weakening in the "conveyor belt" pattern of ocean circulation in the North Atlantic (see p.60) could have significant negative consequences for fisheries in that region.

The good news

Socioeconomic development could partially or even completely offset the negative impacts of climate change on the food supply. There are roughly 820 million people around the globe who are currently undernourished. Taking into account the combined impacts of changing socioeconomic factors and climate change, projections for the number of undernourished people by 2080 range from a reduction to 100 million to an increase to 1300 million.

Somalian famine victims

The line for food in Baidoa, Somalia seems endless. Unfortunately, climate change will only make situations like the devastating famine in Somalia worse.

...Pestilence and death
Human health and infectious disease during times of global climate change

We install heating and air conditioning for indoor climate regulation, build dams for flood control, dig wells to irrigate our fields, and construct dikes to stave off rising seas. But despite our best efforts to insulate ourselves from our natural environment, we remain highly susceptible to climate change, especially when its effects are rapid or unexpected.

Pests and pollen

Of paramount concern is the effect that climate change might have on human health. Diseases can spread as climates change: insects and rodents that carry disease range more widely as climate barriers are lifted. Already there is evidence that vectors such as ticks are spreading to higher latitudes and altitudes in Canada and Sweden. And ragweed is producing more pollen over a longer season as a result of rising temperatures and atmospheric CO_2 concentrations.

Heat can kill

The wake-up call for climate-change-induced human mortality is the European heat wave of 2003 (see p.52). During two extremely hot weeks in early August 2003, nearly 15,000 French people died; across Europe, fatalities approached 35,000. Most of the dead were elderly people who were unable to escape the persistent and oppressive heat. The death rate began climbing several days after temperatures began to rise, and peaked at over 300 additional deaths per day in Paris alone before temperatures finally began to fall. The IPCC report projects an increase in heat-wave incidence with high confidence.

Some like it hot
Mosquitoes and other vectors of disease may spread and flourish as global temperatures increase.

EFFECTS ON HUMAN HEALTH

The IPCC report projects the following climate changes and related health effects in the 21st century.

Predicted climate change (in order of decreasing certainty)	**Anticipated effect on human health**
On land, fewer cold days and nights	Reduced mortality from cold exposure
More frequent heat waves	Increased mortality from heat, especially among the elderly, infirm, young, and those in remote regions
More frequent floods	Increased deaths, injuries, and skin and respiratory disease incidence
More frequent droughts	Food and water shortages; increased incidence of food- and water-borne disease and malnutrition
More frequent strong tropical cyclones	Increased mortality and injury, risk of food- and water-borne disease, and incidence of post-traumatic stress disorder
More extreme high-sea-level events	Increased death and injury during floods; health repercussions of inland migration

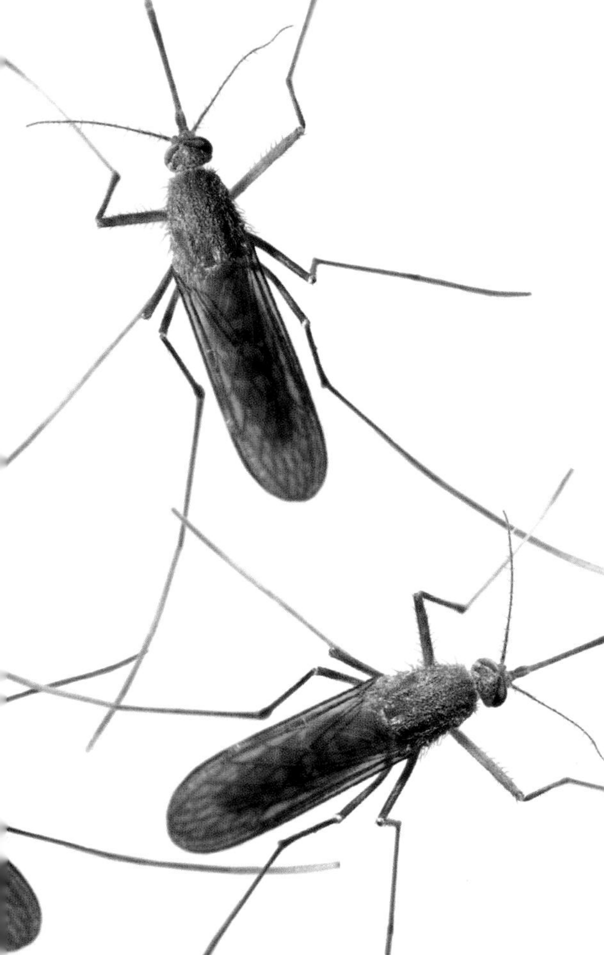

Climate-change-related health impacts will not be uniformly distributed across the world's population. Poor nations will be more susceptible than wealthy ones, because of inadequate access to air conditioning, infrastructure (clean water supplies, electricity, etc.), health care, and emergency response facilities. In all countries, children, the elderly, and the urban poor will suffer disproportionately, as will those people living in low-lying coastal areas.

These threats to human health may serve as a motivating factor for governments to mitigate future climate change. Potential adaptations include raising public awareness, instituting advance warning systems, and improving public health infrastructure in those regions most likely to be hard hit.

Earth, wind, and fire
Impacts on North America

In North America, annual costs associated with weather-related damage already run to tens of billions of dollars. The predicted increase in extreme weather events (see p.100), combined with sea level rise and other climate change impacts, represents a serious threat to people, infrastructure, ecosystems, and the economy.

PATTERN OF TEMPERATURE CHANGE OVER NORTH AMERICA IN RECENT DECADES

The greatest warming is found in the northwest, but all regions have warmed.

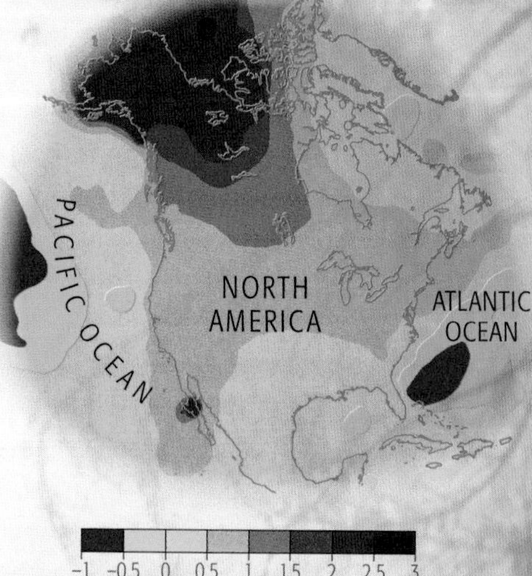

TRENDS IN NORTH AMERICA IMPACTED BY RISING TEMPERATURES

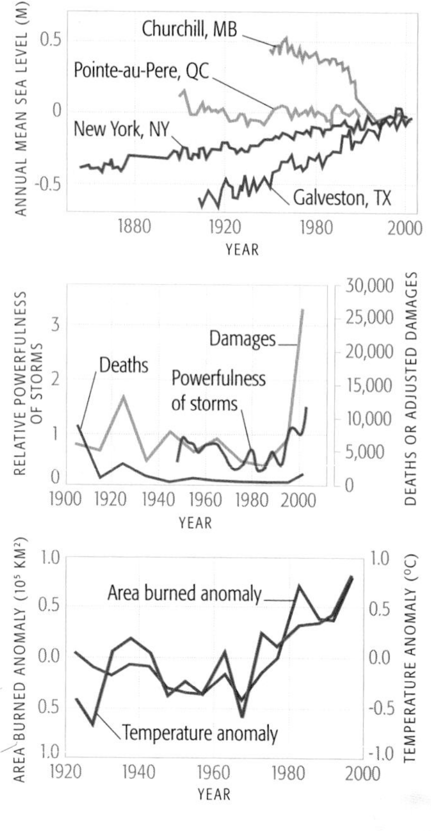

Earth

Changing sea levels threaten to alter coastlines across the globe (see p.98). This graph shows how relative sea levels on the North American coast have changed over the last century. In some regions, such as eastern Canada, the rising of the coastline due to Earth's slow rebound from the last ice age has largely offset the sea level rise resulting from global warming. In other regions, such as the Gulf coast of the United States, this rebound is having the opposite effect and compounding sea level rise.

Wind

Hurricane energy and powerfulness have increased in recent years in the United States as this graph illustrates. It is interesting to note that while damage from storms is increasing – not only due to escalating storm energy and power, but also because of growing coastal populations and more coastal development – mortality hasn't. This is thanks largely to better warning systems and evacuation measures. Still, punishing storms have the potential to wreak havoc, especially on property and infrastructure, in many coastal communities.

Fire

In recent decades in many fire-prone regions, more area has burned as indicated in this graph, which compares trends in surface temperature and forest area burned in Canada over the past 100 years. Warming temperatures mean longer fire seasons, larger forest fires, and a heightened threat to human communities and forest ecosystems.

Earth

The progressive inundation, storm-surge flooding, and shoreline erosion associated with rising sea level is an increased threat to coastal regions. The impacts will be particularly significant along the Gulf of Mexico, and the southern and mid-Atlantic coastlines of North America. In these areas, local sea level is already rising due to Earth's slow rebound from the ice-weight load of the last ice age. (In the extreme northeast, this rebound is acting in the opposite direction, lowering local sea levels.) Rapidly growing coastal populations and overburdened infrastructure make these regions highly vulnerable to the impacts of climate change on property and life.

Wind

Tropical storms are likely to be more powerful along the Gulf and Atlantic coasts (see p.56), where the coastal ecosystems, populations, and infrastructure are already threatened by sea level rise.

Fire

Wildfires in North America have become more common in recent decades, and their incidence is likely to rise still further in response to climate change impacts such as increased drought, decreased winter snow-pack (the amount of winter snow that accumulates at higher elevations), and earlier melt (leading to a longer fire season). For mid-range climate change scenarios, it is predicted that the forest burning area in Canada could roughly double by 2100. Burgeoning populations in regions near the boundary between urban and forested areas will increase human vulnerability to wildfire.

The total area affected by forest fires in the western US has increased by more than a factor of six in the past two decades.

The wider world

Australia and New Zealand may suffer similar impacts to North America, such as intensified droughts and sea level rise. Decreased soil moisture due to warming temperatures will likely lead to increases in the frequency and severity of wildfire in those regions, with negative impacts on agriculture and forestry. As in North America, the significant expansion of infrastructure and population in coastal regions (e.g., Cairns and southeast Queensland in Australia, and the coastline from Northland to the Bay of Plenty in New Zealand) will increase society's vulnerability to rising sea level and (potentially more powerful) tropical storms in future decades.

Africa, Europe, and Asia, and some island nations also have coastal populations that are threatened by sea level rise and more frequent and powerful tropical cyclones. Drier summers over Africa, large parts of Europe and the Mediterranean, and the Amazon basin may lead to increased fire risk as well.

Flames from a forest fire
This fire occurred along the Dalton Highway in Alaska. Forests in the Pacific Northwest are increasingly at risk as a result of longer and hotter fire seasons.

Too wet and too hot
Impacts on Europe

The impacts of climate change are already apparent across Europe, from the record-setting heat wave of summer 2003 (see p.52) and the melting of long-standing mountain glaciers, to shifting precipitation patterns and readily observable changes in ecosystems. Some of these changes can be attributed to an atmospheric phenomenon known as the North Atlantic Oscillation. The increasing tendency for the North Atlantic Oscillation to be in its positive phase has favored a stronger, more northerly jet stream over Europe. This may, at least to some extent, be associated with climate change.

Too wet

Average annual rainfall has already increased by nearly 50% over parts of northern Europe. Further increases in winter flooding in coastal regions, and an increased frequency and intensity of flash floods for much of Europe are projected with further warming. Almost 2 million people in the low-lying countries of the Netherlands, Belgium, and Luxembourg could be threatened with flooding in the coming decades as a result of these changes combined with the impacts of rising sea levels.

SELECTED POTENTIAL CLIMATE CHANGE IMPACTS IN EUROPE

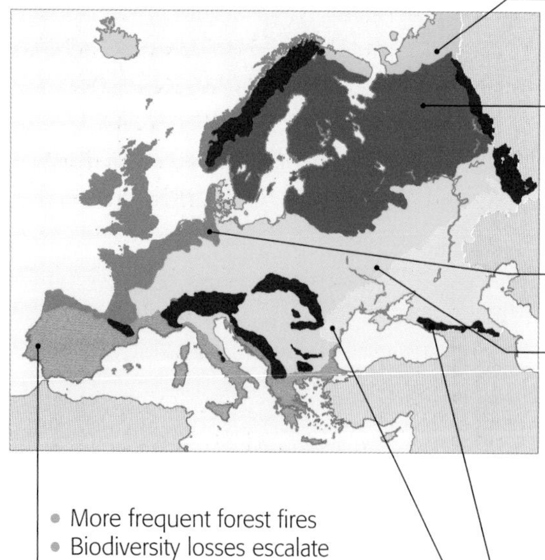

- Thawing of permafrost
- Substantial loss of tundra biome
- More coastal erosion and flooding

- More coastal flooding and erosion
- Greater winter storm risk
- Shorter ski season

- More coastal erosion and flooding
- Stressing of marine ecosystems and habitat loss
- Increased tourism pressure on coasts
- Greater winter storm risk and vulnerability of transport to winds

- Increased frequency and magnitude of winter floods
- Heightened health threat from heat waves

- Severe fires in drained peatland
- Disappearance of glaciers
- Shorter snow-cover period
- Upward shift of tree line
- Severe biodiversity losses
- Shorter ski season
- More frequent rock slides

- Decreased crop yield
- More soil erosion
- Increased salinity of inland seas

- More frequent forest fires
- Biodiversity losses escalate
- Negative impact on summer tourism
- Heat wave impacts grow more serious
- Cropland losses as well as losses of lands in estuaries and deltas

The average rainfall has already increased by 50% over parts of northern Europe.

Too hot

For Europe, the past decade has been the warmest on record for at least the last 500 years (see p.52). Climate model projections indicate that further warming will be concentrated in northern Europe in winter and in southern and central Europe in summer.

It is possible that certain impacts of these changes could be beneficial for human inhabitants. For example, warmer winters could reduce the number of deaths arising from exposure to extreme cold. On balance, however, it appears very probable that the risks to human health will increase. Deadly heat-stress, associated with events such as the 2003 heat wave, will almost certainly become more common in Europe. Increased flooding and warmer winters are also likely to facilitate the spread of water-born, vector-born, and food-born diseases (see p.132).

Across the globe

Already common, heat waves are likely to increase in frequency and intensity over the next few decades as large parts of North America, Asia, Australia, and Africa continue to warm. In the world's cities, urban "heat island" effects, poor air quality, population growth, and an aging population are likely to magnify the impacts of rising temperatures.

Over large parts of North America, Asia, Australia, New Zealand, and South America (e.g., Venezuela and Argentina) increases in heavy rainfall have already led to a higher incidence of floods and landslides. Further increases in such events will likely lead to degraded water quality in these regions (see p.122) and the spread of water-borne diseases.

HUNGARIAN RAINSTORM
This torrential scene took place on Andrassy Street in Budapest, Hungary, in April 2006. More frequent and heavier rainfall can be expected over large parts of Europe as a result of climate change.

The polar meltdown

Slip sliding away
Coastal erosion is a growing problem in the Arctic. The muddy permafrost coastline on the Beaufort Sea of northern Alaska is eroding as waves undercut the melting soil. For scale, the cliff is about 3–4 m high.

Each decade the snow-free period in Arctic Eurasia and North America increases by five or six days, exposing dark ground that absorbs sunlight more effectively than snow. At the same time, the Arctic Ocean's sea-ice cover is decreasing at 7% per decade, exposing the darker sea surface so that the Arctic warms even more. These and other positive feedback loops (see p.24) are amplifying the polar response to the buildup of fossil-fuel CO_2 (see p.92). On average, the Arctic is warming at twice the rate of the globe as a whole, and the region's vast land masses are warming at five times the global average. The situation is less clear in the Antarctic, although a recent study has shown that the Antarctic ice sheet seems to be shrinking for the first time. Continued observation will determine if this really is a downward trend, or simply part of a cycle of shrinkage and growth.

Permafrost thaw

Beneath the Arctic's active layer of soil, which thaws each year, is permafrost—permanently frozen soil. Permafrost is important to human settlement in the Arctic. It has provided a solid and impervious substrate to support building foundations, roadbeds, and pipelines, and to contain sewage ponds and landfill leachate (water that collects contaminants as it trickles through waste).

As a result of recent climate warming and locally released heat from buildings and pipelines, permafrost is now melting. Thawing leads to building collapse, pipeline breakage, roadway degradation, and contamination of surrounding environments. Moreover, methane—a strong greenhouse gas trapped within the permafrost for millennia—is now being released into the atmosphere, further exacerbating warming. By 2050, if current trends continue, the active soil layer will have thickened by 15–50% at the expense of the permafrost below

Meltdown pros and cons

The meltdown may have some benefits to society, such as lower heating costs, greater opportunities for agriculture and forestry, increased river flows (supporting hydroelectric power generation and improved navigation through the Arctic), and easier extraction and transportation of marine resources (including potentially huge hydrocarbon reserves below the Arctic Ocean). However, sea-ice retreat compounded by the rising sea level

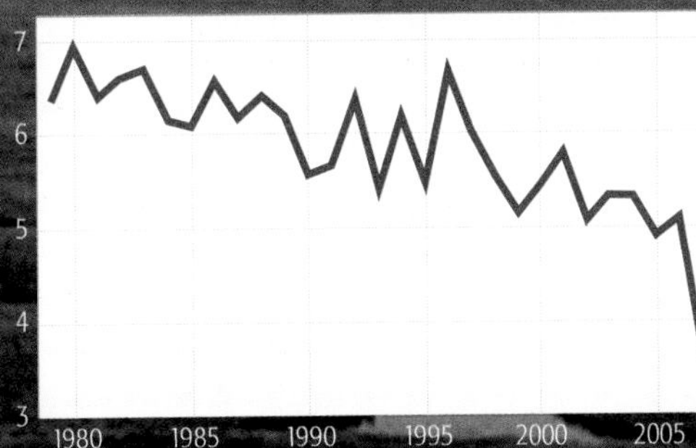

Pacific Ocean
Arctic Ocean
Atlantic Ocean

Minimum extent of sea-ice

- September 1980
- September 1987
- September 1997
- September 2007

SEA-ICE COVER

Satellite observations reveal that the seasonal minimum sea-ice cover in the Arctic, which has been declining for decades, fell abruptly in 2007, reaching a record low. Models didn't predict this level to be reached for several decades to come. One possibility is that the models do not take into account important feedback loops that are amplifying the sea-ice response to warming; this raises concerns that our projections for the future are too conservative, and that sea-ice might soon disappear entirely from the Arctic. It is also possible, however, that the 2007 record low was a temporary downward fluctuation, and that there will be a recovery. Climate scientists will be keeping a close eye on what happens over the next few years.

will also lead to coastal erosion and the collapse of roads, buildings, and pipelines. These facilities—and even whole villages and towns— may need to be relocated. Traditionally, indigenous Arctic people lived in small, widespread communities that could easily relocate when necesary. Modern settlement trends are different: now two-thirds of the Arctic population lives in larger settlements of more than 5,000 inhabitants. This concentrated settlement pattern, combined with a greater dependency on infrastructure, reduces the resilience of Arctic populations to environmental change.

Polar politics

The Arctic has garnered much political attention in recent decades; large, untapped reserves of fossil fuels, minerals, and diamonds await better access for exploration and exploitation. Already, the Russian Federation has sent a submersible to the depths of the Arctic Ocean to plant a miniature flag, and the Danes assert that the geologic feature that traps petroleum in the Arctic Ocean is an extension of Denmark. On a more altruistic note, the developed world is also beginning to recognize and express concerns that the pollution it produces is ending up in the Arctic, with deleterious health and environmental impacts.

The Arctic is warming at twice the rate of the globe as a whole

Part 4
Vulnerability and Adaptation to Climate Change

To reduce our vulnerability to climate change, it is necessary that we both adapt to the effects of the existing buildup of atmospheric CO_2 and reduce the amount of CO_2 we are emitting. In the initial stages of climate change, adaptations can help to allay the threat of rising sea level, diminished and shifting fresh water resources, loss of agricultural productivity, and adverse economic effects. Efforts to reduce or mitigate emissions face many obstacles, but ultimately they may be our best hope for the future, when adaptation alone will be insufficient to counter the long-term impacts of climate change on society, the environment, and the economy.

Is global warming the last straw for vulnerable ecosystems?

How human activity has changed the rules of the game

Temperature has been fluctuating on Earth since the planet developed an atmosphere. So, you might say, why is this such a big problem all of a sudden? We know that the increasing ocean temperatures and atmospheric CO_2 levels linked to coral stress and mortality (see p.114) have geological precedence. Carbon dioxide levels have been fluctuating by 100 ppm for the last million years, and climates have changed rapidly as a result, yet corals have survived. Surely, given their past record of adaptation and survival, corals and other organisms will adapt to global warming?

Not necessarily. This "historical precedence" argument has several weaknesses, not the least of which is the fact that fossil-fuel carbon reserves have the capacity to boost atmospheric carbon dioxide levels further and faster than ever before. There also are key differences between the modern adaptation "playing field" and that of the geologic past. Human land use, construction, pollution, and other constraints make adaptation and survival a different game than it used to be. For example, reefs today are under considerable additional stress from human activity such as:

- Wastewater and sediment discharge
- Pesticides
- Ship groundings
- Dynamiting and poisoning for fish collection
- Increasing recreational use
- In the Caribbean, a possible increase in pathogens carried by dust storms that have been exacerbated by human-caused desertification of Africa's Sahel region

These modern factors have increased the sensitivity of ecosystems to global warming, and compromised their ability to "bounce back" from adversity.

In assessing the vulnerability of ecosystems to climate change we must consider:

- The potential rate of change
- Additional stresses imposed by human activity
- Barriers to adaptation and migration imposed by human activity, human settlement, and infrastructure (e.g., roads and pipelines)

Ecosystems in jeopardy

Will we act to reduce the stresses on ecosystems in advance of significant impact? Will we attempt to restore destroyed habitats or create new habitats before changes are irreversible? Are such efforts likely to be successful? There is much that we do not understand about how ecosystems function and what services they provide (see p.112). But many think that if we don't act now, the anticipated stress of climate change will be "the straw that broke the camel's back."

What is the best course for the coming century?

Up to 1 m of sea level rise could take place by 2100, given "middle of the road" future emissions scenarios (see p.88).

To adapt or to mitigate?

It seems that we have two options: adapt to these changes or mitigate against them. We can take actions to reduce the buildup of carbon dioxide in the atmosphere (mitigation) or to offset the effects of this buildup (adaptation). The problem is that the magnitude of the potential climate change exceeds the capacity for societies and ecosystems to adapt. Moreover, these changes also probably exceed our capacity to mitigate against them, at least for the next few decades (see part 5). Such realities suggest that the best approach is to adopt a plan for the future that is a blend of both adaptation and mitigation, accompanied by technological development to support these efforts and research to guide the technology.

The IPCC report describes the vulnerabilities of countries to climate change with and without mitigation efforts (see maps below and opposite).

Adapt

The vulnerability of each system (i.e., country or region) is related to its adaptive capacity—that is, its "ability or potential to respond successfully to climate variability and change." China and Africa currently

CLIMATE CHANGE VULNERABILITY IN 2100

Vulnerability to climate change can be lessened if mitigation efforts are made and adaptive capacities are enhanced.

KEY (Vulnerability level)
10 Extreme
9 Severe
8 Serious
7 Moderate
6 Moderate
5 Modest
4 Modest
3 Little
No Data

Assuming current adaptive capacities and no mitigation

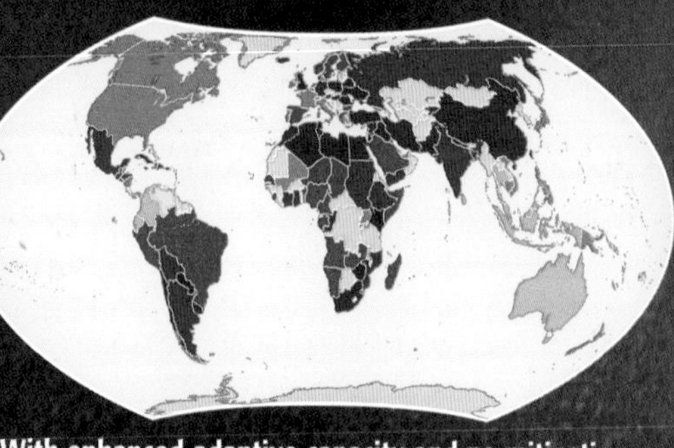

With enhanced adaptive capacity and no mitigation

have higher vulnerabilities than developed nations—and will continue to for the next few decades—because of their low adaptive capacity. Even though the adaptive capacities of developing nations are expected to improve with time, overall vulnerabilities are still predicted to remain high. In fact by 2100, even the nations in the developed world, including the US, could be overwhelmed unless steps are taken to enhance adaptive capabilities and to mitigate against the buildup of CO_2 in the atmosphere.

Mitigate

In the example illustrated by the maps below, mitigation means a stabilization of atmospheric CO_2 at 550 ppm (see p.104). This can realistically be acheived only by significant reductions in fossil-fuel burning rates. Decisive action to mitigate against CO_2 buildup leads to a substantial reduction in vulnerability in all but select regions of Africa, China, and Europe. Mitigation efforts primarily benefit developing countries in the short term (over the next few decades). However, by the end of the century, all nations will benefit from taking such actions, as well as from investments in enhancing their adaptive capacity.

Global action required

To improve their adaptive capacity, nations need to incorporate the likely effects of climate change into their strategies for sustainable development and disaster management. Mitigation efforts deal with global factors (atmospheric CO_2), so they will require international agreements (see p.184). The IPCC report makes it clear that the nations of the world need to take aggressive and immediate action to avoid the looming crisis reflected in the maps shown here.

The road not taken
The nations of the world are making decisions now that will make all the difference in how severe the consequences of climate change will be.

Assuming current adaptive capacities and mitigation steps taken to stabilize atmosphere

With both enhanced adaptive capacity and mitigation steps taken to stabilize atmosphere CO_2

It's the economy, stupid!

Clearly we can reduce the potential damage to natural ecosystems by reducing greenhouse gas emissions. What may not be as apparent is that emission reduction can be good for the economy as well. Economists tell us that the formidable cost of emissions reductions may actually be less than the economic damage that will result from climate change. One fairly conservative estimate links unbridled carbon emissions to economic damages amounting to 2–3% of GDP by the year 2100. Society must balance the cost of these damages against the substantive costs of emission reduction.

The cost of carbon

A so-called "integrated assessment model," which takes into account economic considerations as well as climate change, can be used to estimate the "social cost of carbon" or SCC. This is the cost to society of emitting one additional metric ton (tonne) of carbon. The SCC incorporates the climate change and associated economic impacts of carbon emissions over a prescribed time horizon. This time horizon is typically up to the year 2100, but sometimes it also covers the centuries or millennia over which carbon emissions will continue to affect climate. Estimates of the SCC range from a few US dollars per tonne to several hundred. A prominent economist, William Nordhaus, estimates the present SCC at US $30. In other words, the typical American, who drives 10,000 miles per year and thereby emits a tonne of carbon into the atmosphere, is imposing a cost of US $30 on society. To add insult to injury, this cost impacts society both now and in the future, and the driver is not penalized at all for these damages. Thirty dollars may not seem like much, but think of all the drivers in North America alone, and it really adds up.

According to Nordhaus, emission reductions over the next several decades could save the global economy US $3 trillion.

Carbon credits

Emissions reduction passes a cost-benefit analysis only when the SCC exceeds the costs of carbon reduction—in the example above, that means as long as it costs less than US $30 to offset the emissions of the typical American car driving 10,000 miles per year. The SCC can be used to set the value of carbon credits—credits issued to nations for reducing carbon emissions, e.g., as part of the Kyoto Protocol—or level of taxation. If the SCC is US $30, a 9 cents per gallon gasoline tax would offset the cost incurred by society for the damage done by driving 10,000 miles a year. Nordhaus equates such a tax with other taxes on harmful practices, such as smoking, and contrasts it to taxes on beneficial activities, such as labor.

potential climate damage over the next several decades

$20,000,000,000

minus the potential reduction

$5,000,000,000

plus the cost of reduction

$2,000,000,000

equals net climate damage

$17,000,000,000

Model limitations

Most of the integrated assessment models do not take into account the distinct possibility of abrupt climate change, which could lead to catastrophic damages without historical precedence. They also do not take into account the possibility that climate change will far exceed current predictions. The resulting damages could lead to astronomical increases in the SCC.

Ethical concerns

Finally, it should be noted that ethical concerns may call for action even when cost-benefit analyses do not. The fact that climate change will probably redistribute resources in a "reverse Robin Hood" fashion (see p.130) is particularly unfair to developing nations, such as Bangladesh. The cost of inaction to these communities may be incalculable.

Given this inherently unbalanced scenario, is it fair for the industrial nations—the primary generators of greenhouse gas emissions—to be the ones calling the shots and determining whether or not action is worth taking?

A finger in the dike

Up to 1 m of sea level rise could take place by 2100, given "middle of the road" future emissions scenarios (see p.98). And sea level might ultimately rise as much as 5–10 m if significant parts of the Greenland and West Antarctic ice sheets melt. Even though that could take several centuries to happen, we may be committed to this eventuality by 2100 if emissions meet or exceed the "middle of the road" scenario estimates. Such a substantial rise in sea level would threaten the viability of coastal settlements worldwide (see p.110).

Even a modest future sea level rise would be problematic, given the the current trend towards the aggressive development of coastlines. This impending crisis remains one of the biggest challenges posed by climate change. Adoption of the most austere stabilization policies could reduce the risks of higher-end projections, but we are already likely committed to moderate (i.e., more than 0.3 m) inundation. Some degree of adaptation will be required in addition to mitigation (see part 5).

Rough water

In February 2007, rough seas and gale force winds caused damage to the promenade between Teignmouth and Holcombe in Devon, UK. Faced with rising sea levels, trying to preserve such structures may prove to be a losing battle.

PROTECTING COMMUNITIES FROM RISING WATER

There are three stages of adaptation that coastal communities threatened with rising sea levels may take.

Protection through engineering
The first stage—the most proactive—seeks to protect the population and infrastructure through engineering solutions (e.g., the construction of "empolderings," which structurally reclaim inundated land, and coastal defenses such as dikes or beach nourishments, which create impediments to inundation).

Accommodating inundation
The second stage of adaptation for coastal communities is accommodation. The schemes employed in this stage allow for some degree of inundation (e.g., the building of flood-proof structures, and the use of floating agricultural systems).

Coastal retreat
The third and final stage of adaptation is retreat. Retreat can take various forms (e.g., managed retreat, the building of temporary seawalls, or the monitoring of coastal threat to determine if and when evacuation is necessary).

The cost of inaction

While some adaptation strategies (e.g., the construction of massive coastal defenses, such as sea walls) could be expensive, the cost of inaction is arguably far greater in terms of lost lives and property. And most cost-accounting doesn't include collateral damages to coastal businesses, social institutions, ecosystems, and the environment. For many island and low-lying regions, adaptations are urgently required; in the absence of adaptation, even the climate changes projected in "middle of the road" scenarios could render these regions unlivable. Indeed, one Pacific island (Tuvalu) has already begun to plan for possible future evacuation to New Zealand.

As with other climate change threats (see p.190), ethical considerations arise from the disparity in wealth and resources between developed and developing countries. Adaptation will naturally be more challenging for poorer nations, due to their less-developed adaptive capacity (see p.152), and their limited financial ability to fund costly engineering projects.

Water-management strategies

We know that the global demand for fresh water will rise as population grows. We also know that in many regions, increasing demand will coincide with a decreased water supply owing to the impacts of climate change (see p.122).

The challenges ahead

Current water-management practices are unlikely to be adequate for addressing the new and additional challenges resulting from climate change. How will we alleviate both the stress of worsened water pollution on the environment and ecosystems and the increased flood-risk associated with more intense rainfall? How will we address the repercussions of diminished energy resources resulting from reduced river-flow in many regions (see p.124)? And how will we tackle the problem of dwindling drinking and irrigation stores? Fortunately, there are changes in water-management practices that may help us with the daunting challenges ahead.

Adapting management practices

Communities can commit to making "no-regrets" refinements in water-management practices—that is, changes that will be helpful in dealing with the challenges posed by natural year-to-year variations in climate, regardless of whether or not human-caused global warming ultimately proves to be a threat. For example, in the western US, where there is considerable year-to-year fluctuation in drought and flood conditions due to ENSO (see p.90), existing practices designed to deal with this variability could be exploited and refined to accommodate climate change impacts as well. Examples of adaptation procedures include the development of sea-water desalinization facilities, the expansion of reservoirs

Rolling sprinklers
More efficient irrigation methods may prove to be an effective means of adaptation in the face of diminished fresh water supplies.

and rainwater storage facilities, and improvements in water-use efficiency and agricultural irrigation practices.

Planning for the future

Adaptation strategies are already being developed in regions such as North America, Europe, and the Caribbean in recognition of the potential for changes in precipitation patterns and water availability. In some cases, climate model predictions are being taken into account when adaptation procedures are being designed. However, predictions of regional changes in precipitation and drought patterns are still uncertain (see p.89). So are, consequently, the projected changes in river-flow and water levels. As long as such uncertainties persist, it will remain difficult for water managers to develop optimal strategies. Nevertheless, being ready for more of what we have already seen in terms of year-to-year variability makes good sense, no matter what the future holds.

ADAPTATION OPTIONS FOR WATER SUPPLY AND DEMAND

Supply-side	Demand-side
Prospecting and extraction of groundwater	Improvement of water-use efficiency by recycling water
Increase of storage capacity by building reservoirs and dams	Reduction in water demand for irrigation by changing the cropping calendar, crop mix, irrigation method, and area planted
Desalination of sea water	Reduction in water demand for irrigation by importing products
Expansion of rainwater storage	Adoption of indigenous practices for sustainable water use
Removal of invasive, non-native vegetation from river margins	Expanded use of water markets to reallocate water to highly valued areas
Transport of water to regions where needed	Expanded use of economic incentives, including metering and pricing to encourage water conservation

A hard row to hoe

Climate change impacts on agriculture, livestock, and fisheries may jeopardize our ability to provide adequate food for a growing global population (see p.130). Are there adaptations we can make to protect ourselves from these impacts?

Getting ahead of the curve

Some of our agricultural options include changing crop varieties, locations, and planting schedules in response to changing seasonal temperature and precipitation patterns. These techniques, in some cases, reduce negative impacts and, in other cases, even convert impacts from harmful to beneficial.

Adaptive practices could lead to increased crop yields in temperate latitudes, and potentially maintain current yields in tropical latitudes if warming is only moderate. If warming becomes high enough, however, the growing stress on water supplies may increasingly limit the benefits of adaptative strategies.

While adaptations can offset harmful impacts and even yield positive impacts, implementing them will require both a rethinking of governmental policies and new institutions to facilitate changes at the local level. It is therefore important that these measures be integrated into future economic development strategies.

Adaptation measures are not without some cost, both to communities and the environment. And implementation faces some obstacles. As crop yields begin to decrease in response to climate change impacts, there may be greater pressure on farmers to adopt unsustainable practices in an attempt to maximize short-term yields.

Winners and losers

There are ethical considerations that also come into play. Small farmers and subsistence farmers in tropical regions will be most vulnerable to climate change impacts, due to their relative lack of access to institutions that can facilitate adaptation. Yet, their contribution to greenhouse gas emissions is minimal.

Ironically, the farm industry in temperate regions such as the US, which is a major emissions contributor, may stand to benefit slightly from modest warming (see p.130), and is more likely to have access to any needed aid.

Wheat beats the heat
Wheat yields are projected to increase in the extratropical regions like the US, but to decline in tropical regions. Appropriate adaptations could prevent the latter, as long as future warming levels are moderate.

CLIMATE CHANGE IMPACTS ON AGRICULTURE, LIVESTOCK, AND FISHERIES

Sub-sector	Region	Finding	*Alleviation after adaptation*
+3° to +5°C			
Prices and trade	Global	• Reversal of downward trend in wood prices • Agricultural prices: +10% to +40% • Cereal imports of developing countries: +10% to +40%	
Pastures and livestock	Low latitudes	• Strong production-loss in swine and confined cattle	
Food crops	Low latitudes		• *Maize and wheat yields reduced, regardless of adaptation; adaptation maintains rice yield at current levels*
Pastures and livestock	Semi-arid	• Reduction in animal weight and pasture growth; increased frequency of livestock heat-stress and mortality	
+2° to +3°C			
Food crops	Global		• *550 ppm CO_2 (approx. equal to +2°C with no adaptation) increases rice, wheat, and soybean yields by 17%*
Prices	Global	• Agricultural prices: –10% to +20%	
Food crops	Mid- to high-latitudes		• *Adaptation increases all crop yields above current levels*
Fisheries	Temperate	• Positive effect on trout in winter, negative in summer	
Pastures and livestock	Temperate	• Moderate production-loss in swine and cattle	
Livestock		• Increased frequency of livestock heat-stress	
Food crops	Low latitudes		• *Adaptation maintains yields of all crops above current levels; yields drop below current levels for all crops without adaptation*
+1° to +2°C			
Food crops	Mid-to high-latitudes	• Crop growth less likely to be limited by length of growing seasons • No overall change in rice yield; regional variation is high	• *Adaptation of maize and wheat increases yield by 10–15%*
Pastures and livestock	Temperate	• Livestock grazing less likely to be limited by length of growing seasons; seasonal increased frequency of livestock heat-stress	
Food crops	Low lattitudes	• Without adaptation, wheat and maize yields reduced below current levels; rice yield is unchanged	• *Adaptation of maize, wheat, and rice maintains yields at current levels*
Pasture and livestock	Semi-arid	• No increase in productivity of plant growth; seasonal increased frequency of livestock heat-stress	
Prices	Global	• Agricultural prices: –10% to –30%	

Part 5
Solving Global Warming

Adaptation alone is unlikely to avert the most severe impacts of human-caused climate change. Instead, we must take action to mitigate the buildup of atmospheric greenhouse gases that are responsible for observed and projected global warming. Doing so will require that we reduce our reliance on fossil fuels by altering governmental policies and individual lifestyles. Important first steps we must take include forging cooperative relationships with other nations and rethinking how we, as a global community, can satisfy our energy requirements for key sectors of our economy—such as transportation, buildings, and agriculture—with minimal costs to planet Earth.

Solving global warming

There are two ways to mitigate global warming: we can reduce or eliminate fossil-fuel carbon dioxide emissions, or we can remove the carbon dioxide from the atmosphere (see p.178). When we attempt the latter strategy, and remove CO_2, it is referred to as carbon capture and storage (CCS) or carbon sequestration.

Thankfully, there are no insurmountable technological or scientific reasons why we can't employ either strategy: emission reduction or carbon capture and storage. The only barrier is society itself. Although many countries are attempting to reduce them, emissions continue to grow, and atmospheric CO_2 levels are climbing at rates that exceed previous predictions. This is largely because there are few economic incentives for emission reduction, especially in the two largest emitting nations, the United States and China.

Carbon costs

A potential solution to global warming is to translate the social cost of carbon (introduced on p.146) into a carbon cost that is paid by the consumer who emits. A carbon cost is an amount that consumers must pay (as a tax or as part of an emission permit

For example, if emission of one metric ton of carbon is taxed at US $20, the forestry sector will potentially reduce emissions by just over 1 Gt CO_2 eq. If the carbon cost or tax is raised to $100, the forestry industry may be incentivized to reduce emissions by more than 4 Gt CO_2 eq. (The preferred unit to measure greenhouse gases is the so-called "CO_2 equivalent," which expresses the combined impact of multiple greenhouse gases in terms of the impact of an equivalent amount of CO_2.)

exchange) for the emission of one metric ton (tonne) of CO_2. Taxation, presumably, will not only reduce consumption, but also provide an incentive for the development of non-carbon energy sources.

Emission reduction potential

Mitigation efforts by necessity must span many sectors of the economy, from energy supply, transport, and buildings, to industry, agriculture, forestry, and waste management. The largest potential for emission reductions can be found in some unexpected places, depending on whether we credit reductions at the point of emission or at end-use. If point of emission is credited, then the largest reductions are to be had in the energy supply sector. However, if we look at end-use, then the buildings sector rises in importance (see figures below). In all sectors, emissions are predicted to decrease as the carbon cost increases. Note, though, that with the exception of the forestry sector, investing larger and larger amounts of money in carbon emission reductions leads to smaller incremental gains—the so-called "law of diminishing returns."

In the pages that follow, we will investigate the ways in which emission reductions might be achieved in each of the major economic sectors, and consider in turn alternative "geoengineering fixes," such as carbon sequestration and the reduction of incoming sunlight.

Where do all those emissions come from?

There is no easy fix for the problem of ever-escalating greenhouse gases. Emissions are traced to all sectors of society and the economy. On the following pages, we will discuss the potential for the mitigation of greenhouse emissions in each economic sector, but first let's examine the bigger picture.

The largest contributor to current global greenhouse emissions is the global energy supply sector. Forestry and industry are the next biggest contributors, followed by transport and agriculture. Most emissions are in the form of CO_2, stemming from fossil-fuel burning (see p.160) and deforestation (see p.174). Methane (CH_4) and, to a lesser extent, nitrous oxide (N_2O), which are primarily associated with agriculture (see p.170), are also significant contributors.

Power lines
More likely than not, wherever you see electrical power lines like these, the electricity they carry was originally generated by the burning of fossil fuels.

CO_2 equivalent

In order to make comparisons across sectors, it is important to settle on a unit of measurement that takes into account the differing impact of emissions of different types of greenhouse gases. The preferred unit is the so-called "CO_2 equivalent," which expresses the combined impact of multiple greenhouse gases in terms of the impact of an equivalent amount of CO_2. The CO_2 equivalent is typically measured in either megatons (millions of metric tons) or gigatons (billions of metric tons) of CO_2 (abbreviated as Mt/Gt CO_2 eq).

Who emits?

Although the energy supply sector is currently responsible for the largest emissions (nearly 13 Gt CO_2 eq annually), emissions from other sectors have been increasing as rapidly or more so in recent decades. From 1990 to 2004, energy supply emissions increased by roughly a third, while emissions from forestry increased by nearly a half, largely as a consequence of large-scale tropical deforestation. The developed world is currently responsible for the bulk of worldwide greenhouse gas emissions. However, emission rates are increasing most rapidly in the developing world, reminding us that measures aimed at mitigating greenhouse gas emissions must take into account current trends as well as historical patterns.

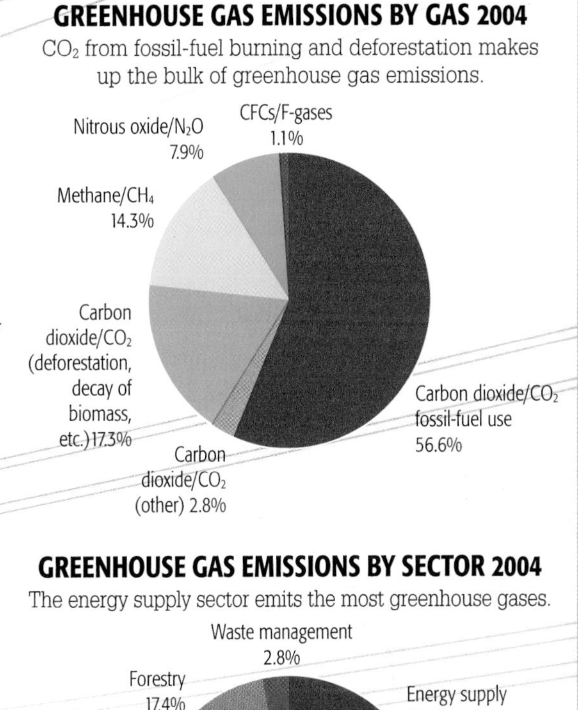

GREENHOUSE GAS EMISSIONS BY SECTOR IN 1990 AND 2004

While the energy supply sector continues to be responsible for the greatest greenhouse gas emissions, emissions from other sectors, such as forestry, are rising at even faster rates.

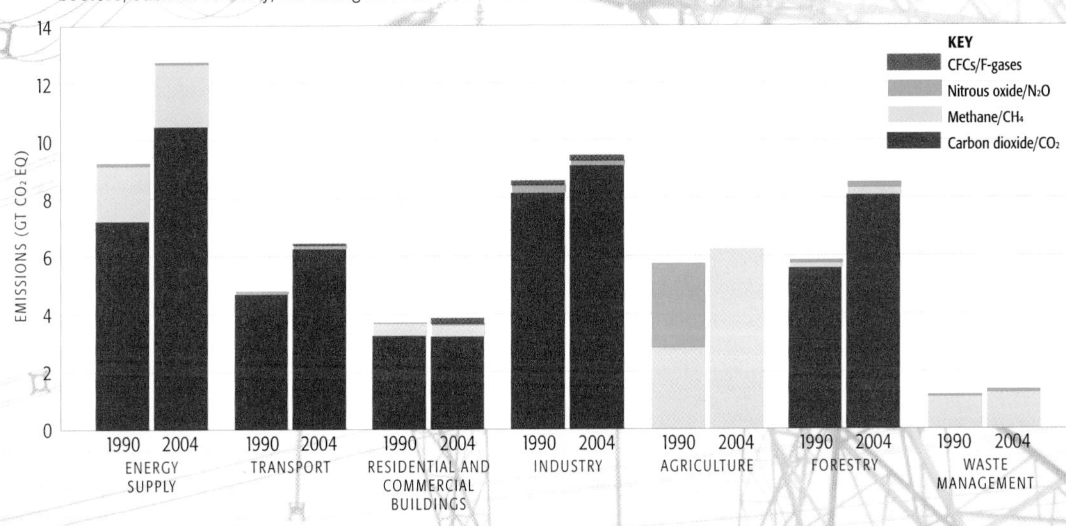

Keeping the power turned on

The combustion of fossil fuels, mainly coal and natural gas, generates much of the world's energy supply—the energy we use for electricity generation and heating (oil is primarily used for transport). The energy sector is the single largest source of greenhouse gas emissions, responsible for over a quarter of all worldwide emissions. The primary culprit is CO_2, though methane released during fossil-fuel processing is also significant. Despite recent international efforts to develop and use non-carbon and renewable energy sources, the introduction of new policies, such as carbon trading, and higher energy prices, emissions have increased substantially in recent years. From just 1990 to 2004, annual energy-related emissions increased from roughly 9 to 13 Gt CO_2 eq.

How can we stabilize emissions?

Without widespread governmental action, energy-related emission rates are projected to rise an additional 50% in the coming decades. As emissions continue and their rate increases, stabilizing greenhouse gas concentrations will become ever more challenging. One common misconception is that the "Peak Oil" phenomenon (the projected impending depletion of readily available petroleum reserves) will solve the fossil-fuel emissions dilemma. However, even if oil wells run dry, the primary sources for the energy sector—coal and natural gas reserves—could last for centuries. In reality, meeting the rising global demand for energy supply while simultaneously slowing the rate of fossil-fuel emissions will require a combination of tools. We need to strive for greater efficiency in power generation, an increased use of carbon-free (e.g., nuclear, solar, and wind) or carbon-neutral (e.g., biofuels) energy sources, and the continued development of carbon capture and storage (CCS) technologies.

Energy alternatives

Carbon-free and carbon-neutral energy sources each have their merits and weaknesses. Increased use of nuclear energy (which currently accounts for about 7% of the global energy supply) is limited by a number of factors, including the restricted availability of uranium, security considerations, safety issues, and limited public support. While renewable energy sources such as solar, wind power, and geothermal are currently minor contributors to the global energy supply, government incentives could encourage development and increased efficiency. However, the localized and variable availability of these sources are obstacles to their widespread use in major urban centers. The use of biofuels—such as wood, sugar cane, vegetable oil, and even (yes) dung for heating and cooking—which currently accounts for more than 10% of global energy consumption, could potentially be modestly expanded (see p.172). The increased use of hydropower, another renewable energy source, faces opposition due to the potential environmental threats posed by major damming projects. In many regions, the viability of hydropower may also be threatened by climate change itself, i.e., shifting precipitation patterns (see p.122).

Wind power
Wind turbine generators turn at Scroby Sands off the coast of Norfolk, UK. The wind farm can be seen from Great Yarmouth.

No easy answers

In short, there is no easy way to meet the world's rising energy demands in a climate-friendly manner. All options need to be taken into consideration, at least in the short term. While the developed world has the highest per-capita energy demand, the most rapid growth in energy use is now taking place in developing countries, such as India and China. Efforts to decrease fossil-fuel emissions therefore will require cooperation between the developed and developing world (see p.184).

WORLD PRIMARY ENERGY CONSUMPTION BY FUEL TYPE IN MEGATONS OF OIL EQUIVALENT (Mt OIL EQUIVALENT)

Despite modest increases in the use of renewable energy resources in recent decades, fossil-fuel sources (gas, coal, and oil) continue to supply the lion's share of the world's energy.

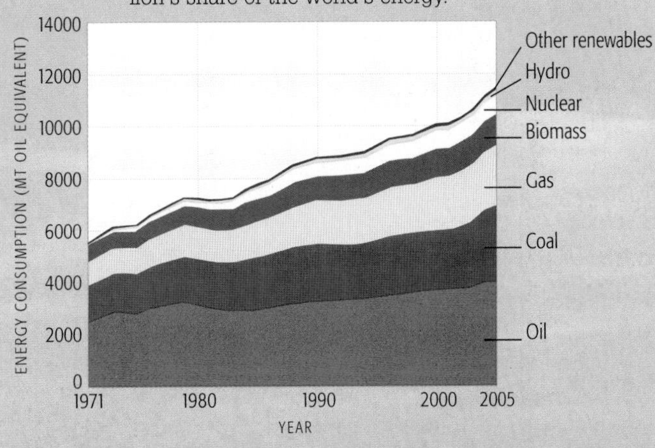

WORLD ENERGY CONSUMPTION BY REGION

While energy consumption is increasing in regions such as China and India, per-capita energy consumption continues to be highest in the developed world.

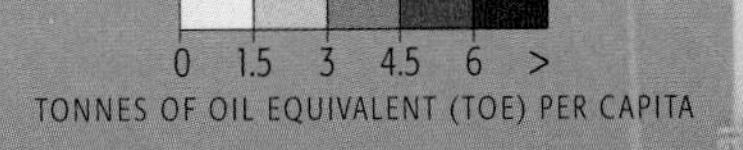

On the road again

Society currently relies almost exclusively on petroleum-based fuels, such as gasoline, for transport. This fuel use results in emissions of about 6 Gt CO_2 eq per year, and it is responsible for 13% of worldwide greenhouse gas emissions. Road vehicles produce the majority (about 75%) of this total. Over the past decade, emissions in the transport sector increased at an even faster rate than those in the energy sector. The greatest growth occurred in the area of freight transport (primarily by trucks for overland freight, and ships and airplanes for international transport). The rate of transport emissions is projected to increase even further over future decades, fueled by continued global economic growth and population increase.

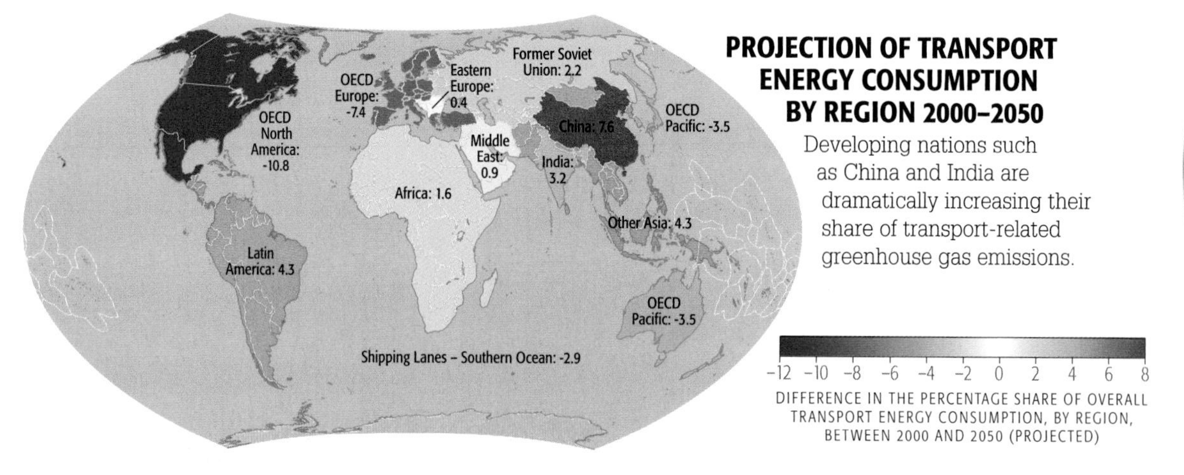

PROJECTION OF TRANSPORT ENERGY CONSUMPTION BY REGION 2000–2050

Developing nations such as China and India are dramatically increasing their share of transport-related greenhouse gas emissions.

HISTORICAL AND PROJECTED TRANSPORT EMISSIONS BY MODE 1970–2050

While land-based modes of transport are likely to continue to dominate transport-related greenhouse gas emissions in the decades ahead, air travel is projected to make an increasingly large contribution.

PROJECTION OF TRANSPORT ENERGY CONSUMPTION BY MODE 2000–2050

If we look at the projected trends more closely, we see that much of the future increase in transport-related energy consumption is likely to come from a combination of personal ("light duty") vehicles and freight trucks.

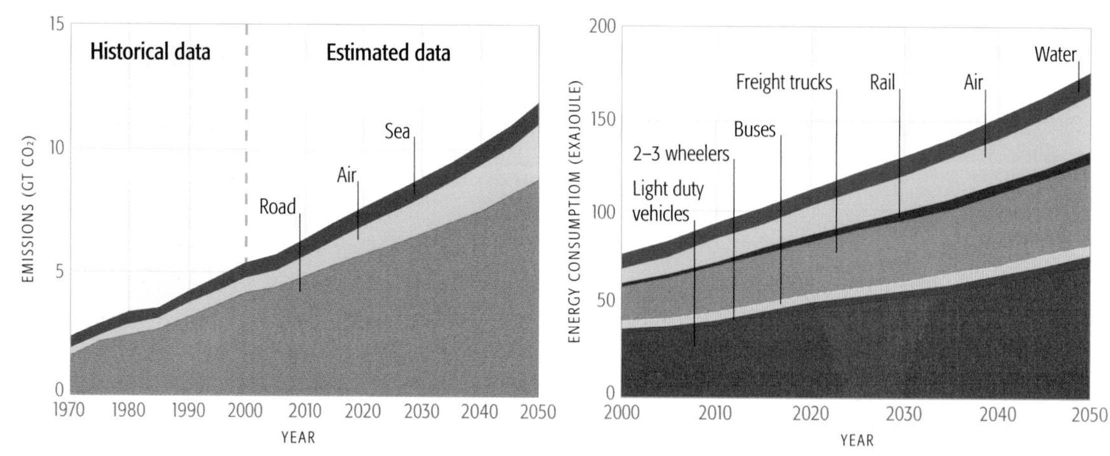

New at the pump
This bioethanol pump at a supermarket in Norwich, UK, serves up biofuels in lieu of fossil fuels. Replacing conventional gasoline with biofuels could reduce transport-related greenhouse gas emissions.

While most people in the developing world still do not possess a personal vehicle, and many have no access at all to motorized transportation, this situation is changing dramatically. In the absence of a radical shift from current practices, transport-related carbon emissions are predicted to nearly double in the next few decades.

The potential of the so-called "Peak Oil" phenomenon (see p.160), to slow the future rate of growth has been greatly overstated. Even if conventional oil fields were to be depleted in coming decades, or drilling were to become prohibitively expensive, sources such as oil shales and tar sands could provide many additional decades of petroleum reserves. As coal liquification technology advances, coal could potentially satisfy the transport sector's rising fuel demands.

Emerging fuel-cell technology has often been cited as a potential solution. However, the use of fuel cells alone for passenger vehicles could simply shift the energy burden from the transport sector to the energy sector, with no net decrease in emissions. This is because fuel cells, which convert fuel energy into electricity in a manner similar to a battery, have to be recharged, and the energy to do that has to come from some place.

The quest for fuel efficiency

Meeting rising global transport sector energy demands while slowing the rate of fossil-fuel emissions will require a combination of greater fuel efficiency and increased use of carbon-free or carbon-neutral technologies. In the short term, this can be accomplished with more fuel-efficient vehicles, such as gasoline/electric hybrid cars and clean diesel vehicles, and through increased use of certain types of biofuels as gasoline additives or substitutes. Two decades from now, biofuels could satisfy 5–10% of the total transport energy demand. Technological innovations and improved air traffic management could result in better fuel efficiency in the aviation sector. Increased reliance on trains, buses, and other public transport, car-pooling, and non-motorized transportation (such as cycling and walking) could also help to curb emissions.

Refueling station for electric vehicles
In an environmentally friendly future, we could see stations like these replacing gas pumps as electric vehicles replace today's inefficient gas-guzzlers.

"No regrets"

In many cases, measures to reduce fossil-fuel consumption constitute "no regrets" strategies. For example, using our automobiles less has the desirable added benefit of reducing traffic congestion and improving air quality. And decreased gasoline consumption results in the national security bonus of reducing our reliance on volatile regimes. Furthermore, improved vehicle efficiency measures lead to savings in fuel expenditures—savings that can be invested in other areas of the economy.

Long-term plans

Additional emerging technologies may allow for further reductions in transport-related greenhouse gas emissions in the coming decades. New types of biofuels, electric and hybrid vehicles with more powerful and longer-running batteries, and more efficient aircraft can all substantially contribute to long-term emission-reduction goals.

Who can help?

Public policy measures can aid the mitigation of transport-related emissions in a number of ways. This is particularly true for nations or communities that are still in the process of establishing transportation systems. For example, well-thought-out urban planning and land-use regulations can provide better access to public transport, and make it more appealing as a commuting option. Thoughtful urban planning can also reduce commuting distances. Governments can establish, enforce, and, where necessary, raise mandatory fuel economy standards. Appropriate taxation and fees can encourage the use of efficient vehicles. Some of the obstacles to the success of these measures are the persistent preference of many consumers for vehicles with poor fuel efficiency, targeted corporate advertising campaigns that reinforce these preferences, and lobbying efforts by car companies to keep fuel efficiency standards low. Any substantive reduction in the transport sector's fuel consumption will require not only proactive government policies, but also greater corporate accountability and personal acceptance of responsibility by individual citizens.

Green Building
This LEED gold-rated academic building on the Pennsylvania State University (USA) campus is both energy efficient and functional.

Building green

The commercial and residential buildings sector is a large emitter of greenhouse gases, accounting for nearly 4 Gt CO_2 eq in 2004. Actually, this sector is also a significant consumer of energy, so if we include this energy use, the buildings sector becomes one the largest emitters of carbon dioxide, emitting approximately 9 Gt CO_2 eq per year in 2004. Including the energy-consumption-related emissions is important as we consider the climate impact of approaches taken to promote energy efficiency in the buildings sector. Happily, many of the approaches taken to reduce emissions in this sector are technologically mature and provide benefits in addition to reducing emissions.

There are two basic ways the buildings sector can reduce its carbon footprint:

- Reducing energy consumption in construction and building operation
- Switching to low-carbon or carbon-free energy sources

Here we focus on the first of these strategies; alternative energy sources are addressed in "Keeping the power turned on" (see p.160).

Reducing energy consumption

The green building movement encourages efficiency in the design, construction, operation, and demolition of buildings, with the goal of enhanced human health and reduced impact on the environment. In the United States, this movement is exemplified by Leadership in Energy and Environmental Design (LEED). Certification points are awarded by LEED for sustainability and efficiency, as well as for the optimization of energy performance and the use and re-use of recyclable materials. A successful LEED building takes into account such important concerns as the availability of alternative transportation to the building (e.g., buses, trains), habitat preservation, and indoor environmental quality.

Reducing energy use in new buildings means reducing the heating and cooling loads. This can be accomplished through passive solar design (taking best advantage of available solar energy) and better insulation. High-efficiency lighting, appliances, and heating and cooling systems, and high-reflectivity building materials, and multiple glazing in windows can also markedly reduce a building's emissions.

Green renovation

Energy savings in new construction can exceed 75%, which bodes well for the future. However, buildings have long lifetimes, so most buildings in existence today will still be in use in 2030. This means that close attention has to be paid to the renovation of existing buildings, because that is where most buildings-related emissions reductions will be made.

Although these so-called "green" renovation techniques can involve considerable costs up-front, there are economic savings in the long term associated with energy-use reduction, and a number of co-benefits as well, including improved indoor air quality. Green building can also create jobs and new business opportunities, which in turn enhances economic competitiveness and energy security.

To enhance the lure of these long-term benefits, some measure of government intervention may be necessary. Appliance efficiency standards, new building codes, mandatory labeling and certification, energy efficiency quotas, and tax benefits for green construction are all options. Most of these mechanisms have a high cost-effectiveness, and in many cases benefits can be realized without costs.

Stemming the rising greenhouse gas emissions from the buildings sector will require a strong political commitment to green construction. This may take the form of governmental monitoring and the enforcement of codes and regulations. In the end, though, if the green building movement is successful, the interior space where we spend much of our time will be healthier and more comfortable, and have a minimal impact on the global environment.

Industrial CO_2 pollution

The image of a factory belching out smoke is engrained in our minds as the epitome of environmental pollution. Although many industries have taken significant steps to reduce pollution, thanks to its intense use of energy, the industry sector is still a major source of carbon dioxide and other greenhouse gases. And that source is growing: from 6 Gt CO_2 eq in 1971 to nearly 10 Gt CO_2 eq in 2004. Of these 10 Gt, approximately 5 Gt are from heat and power production and 5 Gt are from industrial processes, such as cement production and chemical processing. According to the A1B "middle of the road" fossil-fuel emissions scenario (see p.86), there will be little change in CO_2 equivalent emissions from the industrial sector by 2030.

CARBON CAPTURE AND STORAGE (CCS)

Carbon dioxide can be captured before it is released into the atmosphere and transferred underground via pipeline. Possible repositories include coal and salt beds, depleted oil and gas reservoirs, and saline aquifers.

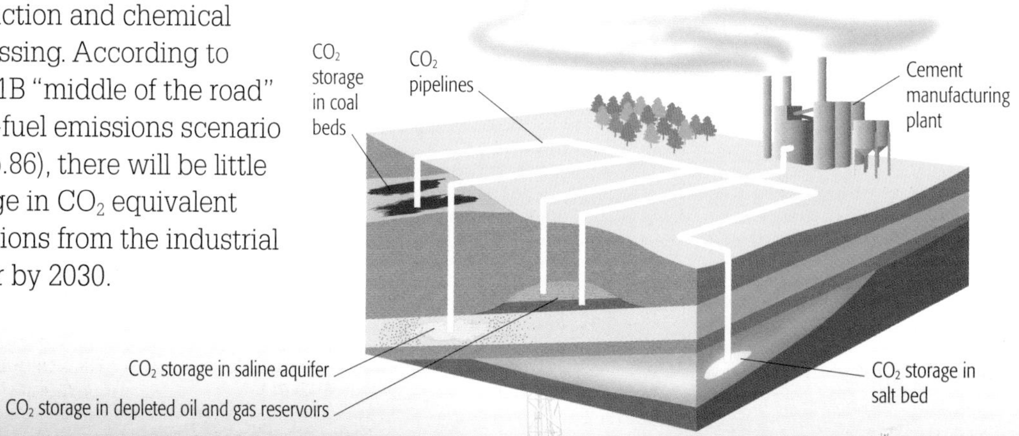

CO_2 storage in coal beds
CO_2 pipelines
Cement manufacturing plant
CO_2 storage in saline aquifer
CO_2 storage in depleted oil and gas reservoirs
CO_2 storage in salt bed

Back to where it came from

The Norwegian oil company Statoil is injecting about 1 Mt of carbon dioxide each year a kilometer below the seafloor at its Sleipner West field in the North Sea.

Industrial CO_2 emission is increasingly becoming an issue for developing nations. In 1971 only 18% of industry-related CO_2 emissions came from developing countries, but by 2004 their share had risen to 53%. This shift reflects both the growth of industry in developing countries and the movement toward improved energy efficiency in developed countries.

Industrial mitigation

There are numerous opportunities for mitigation of greenhouse gas emissions in industry, in part because many factories are using old and inefficient processes. Retrofitting of these factories—replacing electric motors and boilers, using recycled materials for fuel, and fixing leaks in furnaces and air and steam lines—could go a long way toward limiting carbon emissions in the future. Carbon capture and storage is another promising strategy to help realize industrial reductions.

REDUCING INDUSTRIAL CO_2 EMISSIONS THROUGH TAXATION & CARBON SEQUESTRATION

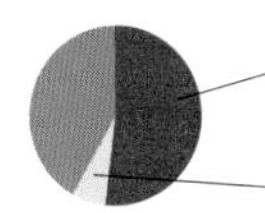

STEEL

Total amount of projected CO_2 emissions by 2030: **1.2 billion metric tons (Gt)**

Tax: \$20–50 per metric ton **Emission reduction:** 0.43–1.5 Gt CO_2/yr

CCS Cost: \$20–30 per metric ton **Emission reduction:** 0.07–0.18 Gt CO_2/yr

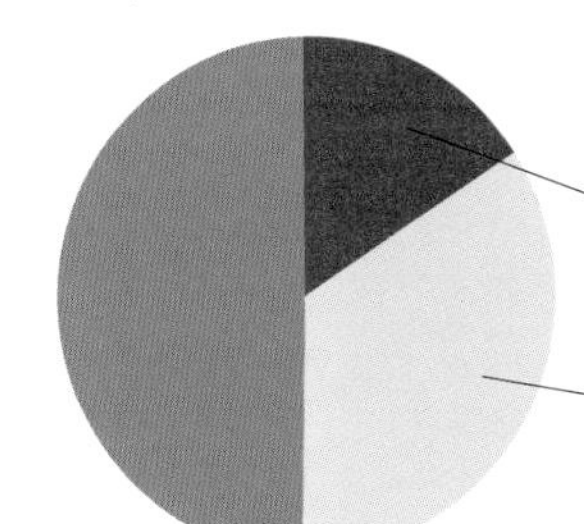

CEMENT

Total amount of projected annual CO_2 emissions by 2030: **6.5 billion metric tons (Gt)**

Tax: up to \$50 per metric ton **Emission reduction:** 0.72–2.1 Gt CO_2/yr

CCS Cost: \$50–250 per metric ton **Emission reduction:** 0.25–4.5 Gt CO_2/yr

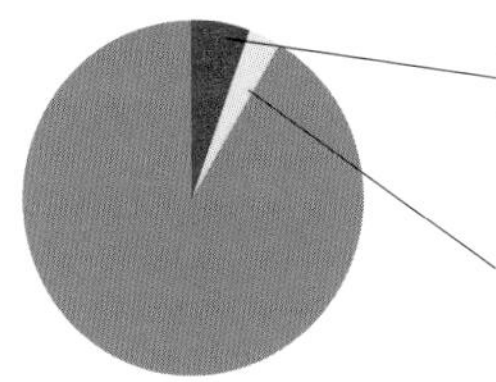

PETROLEUM REFINING

Total amount of projected annual CO_2 emissions by 2030: **4.7 billion metric tons (Gt)**

Tax: up to \$20 per metric ton **Emission reduction:** 0.15–0.30 Gt CO_2/yr

CCS Cost: \$17–31 per metric ton **Emission reduction:** 0.08–0.15 Gt CO_2/yr

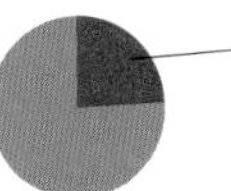

PULP & PAPER

Total amount of projected annual CO_2 emissions by 2030: **1.3 billion metric tons (Gt)**

Tax: up to \$20 per metric ton **Emission reduction:** 0.05–0.42 Gt CO_2/yr

Carbon capture does not apply to this industry

PROJECTED ANNUAL CO_2 EMISSIONS BY 2030
These values include emissions resulting from energy supplied to industry from the energy sector.

EFFECT OF CARBON TAX ON CO_2 EMISSIONS
If government agencies were to impose a tax of the amount listed (US \$ per metric ton of emissions) on an industry, then it is estimated that the taxed industry would potentially reduce its annual emissions by the amount shown in red.

COST OF CARBON SEQUESTRATION
Some industries also have the potential to remove and store CO_2 from industrial emissions before it is released to the atmosphere. Different industries can sequester different amounts per year. If CCS procedures cost the amounts listed above (per metric ton of emissions), then it is estimated that each industry would potentially reduce its emissions by the amounts shown in yellow.

Greener acres

Given that nearly half Earth's land surface is used for farming (crops and grazing), it should come as no surprise that the agriculture sector is a significant contributor to global greenhouse gas emissions. Farming and agriculture are responsible for annual emissions of about 6 Gt CO_2 eq. This is about 13% of worldwide greenhouse gas emissions, and roughly equal to the contribution from transport. Agricultural emissions have increased by 17% over the past 15 years. Much of that increase has come from the developing world, which is now responsible for about 75% of worldwide agricultural emissions. Interestingly, net CO_2 emissions from agriculture are negligible; plants do produce CO_2, but they consume it at about the same rate. The main agricultural emission is methane (CH_4), which is produced by microbes that thrive in environments such as rice paddies and the stomachs of ruminants like cattle, oxen, and sheep. Nitrous oxide (N_2O) from manure and other fertilizers is another agriculture-generated greenhouse gas.

Agricultural mitigation

Carbon is absorbed by and stored in living plants on Earth. Farming and grazing lands thus represent large potential carbon stores (places where carbon is sequestered away from the atmosphere). The land's ability to store carbon has decreased significantly over time, due to heavy-handed agricultural practices. Consequently, the greatest potential for agricultural mitigation lies not in the reduction of emissions themselves, but in the improved management of agricultural lands, which will restore their ability to sequester CO_2. Positive management practices include reducing soil tillage and restoring carbon-absorbing organic soils. More Earth-friendly practices that restore degraded land, such as converting aging crop land to grassland, can similarly aid in mitigation.

Of course, it is also a good idea to decrease agricultural emissions. Improved management practices can play a key role in this endeavor. For example, more efficient fertilizer delivery methods can minimize nitrous oxide emissions. Rice paddies can likewise be better managed to reduce methane production. And using alternative feeds will result in less methane production by ruminants. For example, recent experiments show that adding certain types of food-industry byproducts, such as cooking fat, to cattle feed reduced their methane production.

GLOBAL MITIGATION POTENTIAL OF VARIOUS AGRICULTURAL MANAGEMENT PRACTICES BY 2030

Better management of crop lands, grazing lands, and soils can decrease net greenhouse emissions by allowing land to more effectively sequester atmospheric carbon dioxide. Better farming practices can also reduce methane production, another important contributor to greenhouse gas emissions. The negative values on the graph indicate that rather than mitigating, the action in question is adding to current emissions.

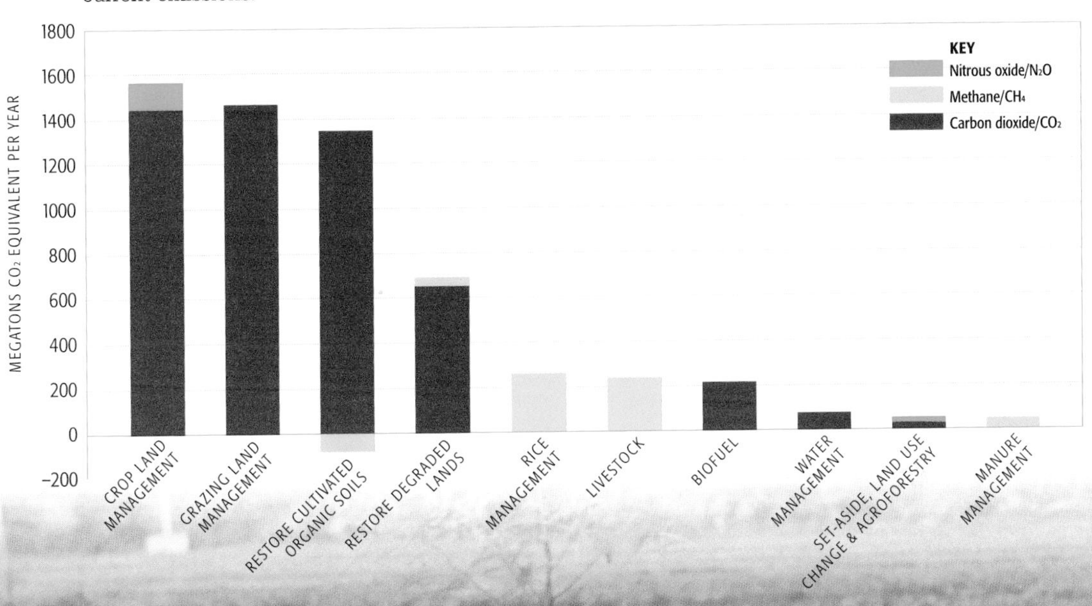

Gathering rice husks

If farmers, like these in southern India, take good care of their farm lands, they can dramatically increase the land's ability to sequester CO_2 from the atmosphere.

Biofuel options

There is also potential for mitigation in agricultural production of biofuels. Burning biofuels does not lead to any net increase in greenhouse gas concentrations because the carbon released was just recently removed from the atmosphere and has only been stored in plants for a few months. For every gram of CO_2 released when a biofuel is burned, a gram was removed from the atmosphere by photosynthesis just a short time ago. This balance is why biofuels are considered "carbon neutral." Crops and agricultural residues can be used either as crude biofuel energy sources (e.g., burned for heat), or they can be chemically altered to yield more efficient biofuels such as ethanol or biodiesel.

Corn, one of the most widely grown cereal crops, can be readily converted into ethanol. However, there are at least two problems that limit prospects for the widespread use of corn-based ethanol as a fuel. First, there are troubling ethical considerations associated with the prospect of trading food for energy in this way, when starvation and malnutrition still afflict large numbers of people, especially in developing countries. Secondly, and perhaps more practically, the processes used to convert corn to ethanol are not very efficient; they require a considerable input of energy, limiting net gains. Researchers are currently seeking alternative pathways for ethanol production. So-called "cellulosic ethanol," which can be derived from agricultural products such as switchgrass, could yield larger net gains in the future. (Switchgrass is a tall grass that grows naturally on the prairies of North America. Research has demonstrated that it has the potential to yield biofuel more efficiently than corn. In comparison with corn, switchgrass is more hardy with respect to soil and climate conditions, is perennial, and requires far less fertilizer and herbicide.)

Uncertainties

Most (approximately 70%) of the potential for mitigation in the agricultural sector lies in the developing world, specifically in China, India, and much of South America. It is challenging to determine precisely what the mitigation potentials of various options are. Key uncertainties include the willingness of governments to promote and support mitigation practices, and the willingness of individual farmers to adopt preferable management practices. Also uncertain is the continued effectiveness of various mitigation strategies in the face of escalating climate change, growing populations, and evolving technology.

Corn growing in Minnesota
Corn, which can be made into ethanol, is one possible source of biofuels. Other alternative biofuels, however, may yield greater mitigation benefits.

ESTIMATED MITIGATION POTENTIAL IN THE AGRICULTURAL SECTOR BY 2030

Note that South America, China, and India have the greatest mitigation potential.

Forests
Source or sink for atmospheric CO_2?

Long before humans were burning fossil fuels, we were contributing to the buildup of carbon dioxide in the atmosphere. The practice of deforestation has accompanied human settlement and agriculture across the globe. The combustion of timber for energy and the gradual decay of lumber used in construction both release CO_2 into the atmosphere. In 2004, the forestry sector emitted roughly 17% of the total greenhouse gases released to the atmosphere.

Carbon uptake

During the pre-industrial era, forest-clearing and wood burning were common practices in Europe and the United States. Recently, however, previously cleared agricultural lands have returned to forests in these regions. As a result, forest-related CO_2 emissions have declined over the last several decades (see graphs opposite) and reforested lands have now become carbon sinks.

Now the developing world is repeating history, aggressively cutting down and burning trees (see map opposite). Deforestation in tropical South and Southeast Asia, Africa, and South America has recently accelerated. Between 2000 and 2005, an area roughly the size of Ireland was lost each year to deforestation.

This new carbon influx from tropical deforestation has been partially offset by the reforestation uptake in Europe, America, and elsewhere. Current emissions from deforestation amount to nearly 6 Gt CO_2 eq per year. If we include emissions from decomposition of logging debris, peat fires, and peat decay, forestry emissions add up to over 8 Gt CO_2 eq per year. On the flip side, it is estimated that approximately 3.3 Gt CO_2 eq per year are taken up through reforestation.

Obviously, the most expeditious way to reduce CO_2 emissions from the forestry sector is to prevent deforestation. But the developed world must keep in mind its own history.

Reforestation potential

What about mitigating via reforestation efforts? The efficacy of reforestation is contingent on favorable climate conditions. Deforested soils tend to dry out, and they are often low in nutrients because most of the ecosystem nutrients were stored in the harvested trees. For this and other reasons, reforestation of tropical rainforest has often been unsuccessful, and many experts believe that reforestation may not be a significant carbon sink in the second half of this century.

Timber!
This Indonesian tropical rainforest is being clear cut so that a palm-oil plantation can be established.

RATE OF CHANGE IN FORESTED AREA

This map shows the rate of change in forested area between 2000 and 2005. Note that the highest rates of deforestation (in red) are largely in the tropics.

< -0.5 0.5 >

NET CHANGES IN FORESTED AREA BETWEEN 2000 AND 2005 (PERCENTAGE CHANGE PER YEAR)

HISTORICAL TRENDS IN FOREST CARBON EMISSIONS AND UPTAKE

These graphs show historical trends in forest carbon emissions (red) and uptake (green), for the period between 1855 and 2000, in Mt CO_2 eq. The US and Europe have become net carbon sinks after a long history of deforestation.

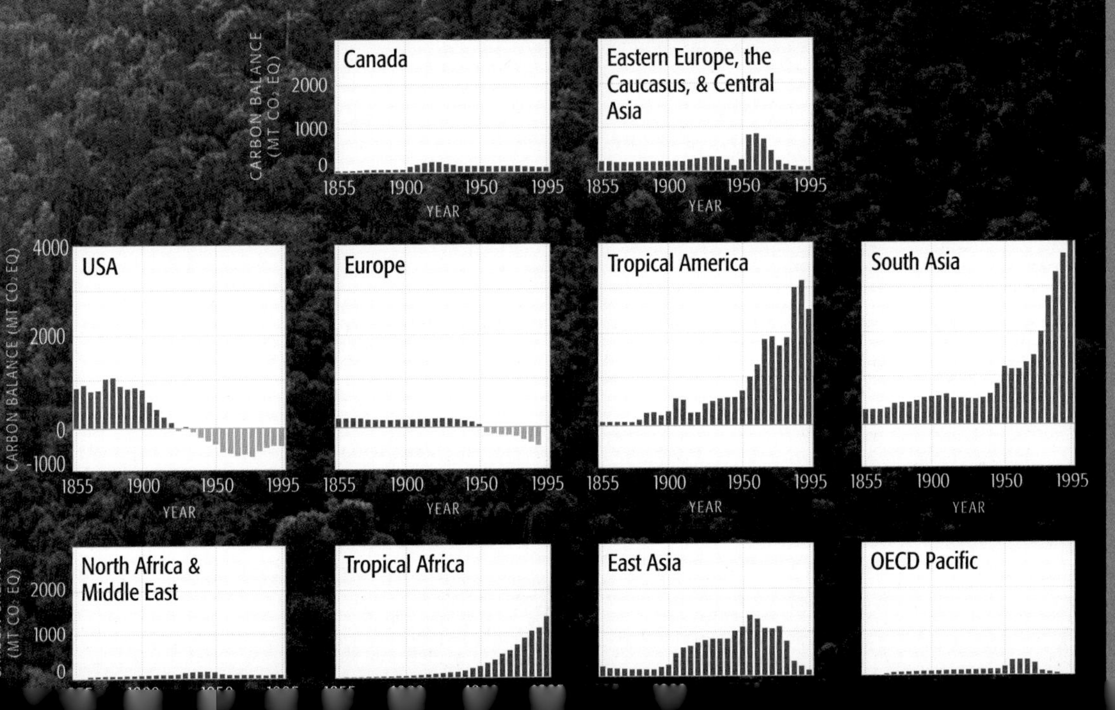

Waste

Life pollutes: humans pollute, cats pollute, ferns pollute, and bacteria pollute. Organisms metabolize food to gain the energy they need to grow, reproduce, and move. Metabolism creates waste products; these are often toxic and need to be eliminated. Although some level of pollution is unavoidable, humans pollute for reasons other than just simple metabolism.

Landfill problems

Although waste disposal is a significant problem facing an ever-expanding and consumptive world population, greenhouse gas emissions from the waste management sector account for only about 3% of total global emissions (1.3 Gt CO_2 eq per year). The largest share of waste-related emissions comes from landfills, primarily in the form of methane. The bacteria that decompose waste in the oxygen-depleted interior of landfills produce methane. Well-aerated landfills produce carbon dioxide instead of methane, which molecule-for-molecule is a weaker greenhouse gas.

Researchers digging in landfills have exhumed decades-old hot dogs that showed no signs of decomposition. In the context of the carbon cycle, those hot dogs, and much of the other waste accumulating in typical, inefficient landfills, are carbon "sinks." To some extent carbon sinks offset the release of carbon dioxide from fossil-fuel burning. For example, a corn plant may take up a carbon dioxide molecule recently emitted from a coal-fired power plant. The corn is then fed to a pig, and the pig is slaughtered to make hot dogs. A hot dog is discarded by a child at the ball game, and ends up in a landfill where it may sit for centuries without decomposing and releasing its carbon back into the atmosphere. Taken together, all the hot dogs and the rest of the 0.9 Gt CO_2 eq of waste generated globally each year "store" about 0.2 Gt CO_2 eq. This number is small, even compared to the relatively low total waste sector emissions, so we certainly don't want to produce more waste just to sequester a small amount of carbon.

Is that a kitchen sink or a carbon sink? Landfills like this one in Tokyo Bay, Japan, both accumulate carbon wastes and emit greenhouse gases.

Energy from waste

Rather than attempting to reduce already low waste sector emissions further, some communities are exploring waste recycling alternatives, including burning landfill garbage to use as a renewable energy resource (as a fossil-fuel substitute). Society is already obliged to collect waste and transport it to a central repository; why not burn it for inexpensive energy rather than just allowing it to accumulate and release methane? State-of-the-art technology is able to prevent much of the potential pollution from incineration. The methane seeping out of landfills can also be captured and used for energy production, further reducing the overall impact of landfills on the environment. This large-scale garbage "recycling" may turn out to be a win–win situation for society as we struggle to find ways to mitigate emissions and efficiently manage waste.

WASTE SECTOR EMISSIONS PROJECTION THROUGH 2050, AND THE EFFECT OF VARIOUS MITIGATION STRATEGIES

The emission of greenhouse gases from landfills and other waste-disposal activities will increase dramatically over the next 40–50 years, but increased incineration of waste (to use as a substitute for fossil-fuel energy), and expanded efforts to capture landfill methane can slow the increase.

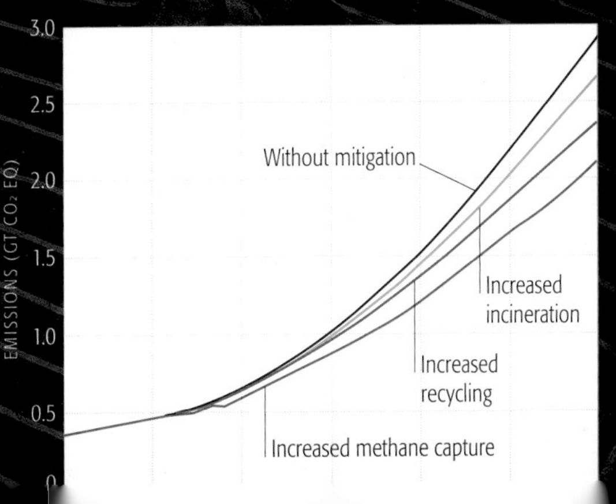

Geoengineering
Having our cake and eating it too

Geoengineering is an alternative approach to mitigation that involves using technology to counteract climate change impacts either at the source level (doing something about growing greenhouse gas levels) or at the impact level (offsetting climate change itself). These approaches involve planetary-scale environmental engineering the likes of which society has never before witnessed.

Carbon sinks

One source-level geoengineering proposal, referred to as "iron fertilization," involves adding iron to the upper ocean. Iron is a nutrient that is of limited availability in the upper ocean. This scarcity of iron places limits on the activity of marine plants that live near the ocean surface. Some scientists think that iron fertilization can increase the rate at which plants in the upper ocean take up carbon dioxide, thus boosting the efficiency of the deep-ocean carbon sink (see p.94), and offsetting the buildup of carbon dioxide in the atmosphere. However, the limited experiments that have been done suggest that the main effect of iron fertilization would probably simply be faster cycling of carbon between the atmosphere and the upper ocean, with little or no burial of carbon in the deep ocean. In addition, there could be possible negative side effects if humans interfere further with the complex and delicate ecology of the marine biosphere. Other geoengineering approaches include attempts to increase the efficiency of terrestrial carbon sinks by planting more trees and "greening" regions that are currently deserts. Many consider this approach far more environmentally friendly than other proposed schemes, but it is unclear if it could be accomplished on the massive, planetary scale required to significantly offset human carbon emissions.

Carbon capture

Closely related to regional greening plans are so-called carbon capture and sequestration (CCS) approaches. In CCS approaches, carbon is extracted from fossil fuels as they are burned, preventing its escape and buildup in the atmosphere. The captured carbon is then buried and trapped well beneath Earth's surface or injected into the deep ocean, where it is likely to reside for many centuries. One potentially effective CCS scheme would involve scrubbing carbon dioxide from smokestacks, and reacting it with igneous rocks to form limestone. This mimics the way that nature itself removes carbon dioxide from the atmosphere over geological timescales (see p.94). Recently, Klaus Lachner of Columbia University has argued for a related alternative, in which massive arrays of artificial "trees" take carbon directly out of the air and precipitate it in a form that can be sequestered.

Saltwater in the sky
This artist's conception shows a proposed device for spraying large quantities of sea water into the atmosphere to help boost the sun-reflecting power of marine stratocumulus clouds.

Solar shields and aerosols

One of the most frequently proposed impact-level geoengineering approaches involves deliberately decreasing the amount of sunlight reaching Earth's surface to such a degree that the reduction in incoming radiation offsets any greenhouse warming. One such method involves deploying vast "solar shields" in space that reflect sunlight away from Earth. Shooting sulphate aerosols into the stratosphere to mimic the cooling impact of volcanic eruptions (see p.18) is a less costly, but potentially more dangerous alternative. This method could exacerabate the problem of ozone depletion (see p.30) by tampering with the chemical composition of the stratosphere.

GEOENGINEERING SOLUTIONS TO CLIMATE CHANGE

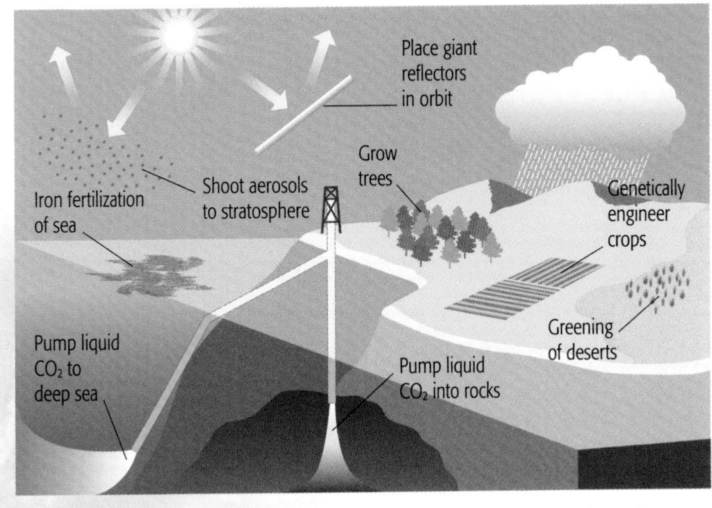

Various geoengineering schemes, such as the ones illustrated here, have been proposed to offset the impacts of fossil-fuel burning on climate.

While calculations suggest that either of these impact-level methods could indeed offset greenhouse warming of the atmosphere, they each come with problems of their own. First, they do nothing to avert the problem of ocean acidification associated with increasing atmospheric carbon dioxide levels (see p.114). Furthemore, climate models indicate that reducing the incoming solar radiation, while potentially offseting the warming of the globe, would not necessarily counteract the regional impacts of greenhouse warming. Some regions might warm at even greater rates, and patterns of rainfall and drought could be dramatically altered. Not to mention that if, for some reason, these activities were ultimately halted, the full impact of warming that had remained hidden for decades would suddenly be unmasked, leading to dramatic, rapid global climate change.

Schemes of last resort

Each of the proposed geoengineering schemes has possible shortcomings and poses a potential danger. Some advocates maintain that if we are backed into a corner and faced with the prospect of irreversible and dangerous climate change, we may need to resort to these schemes at least as partial solutions. Others note that it would be wise not to tamper with a system such as the climate, the workings of which we still do not entirely understand. Either way, the debate over whether geoengineering is likely to be an effective and prudent solution to climate change is bound to continue—as scientists will continue to come up with new proposals for using technology to address climate change problems.

But what can I do about it?

If all this talk about energy sectors and governmental buy-in leaves you feeling helpless in the face of global warming, don't let it! We can all make lifestyle choices that will directly aid in the mitigation of greenhouse gas emissions. In many cases, these are "no regrets" changes that have positive side benefits, such as improving our health and quality of life, conserving natural resources, and facilitating greater environmental sustainability.

Lifestyle choices

First and foremost, we can be more efficient in our use of energy. We can save energy daily by making home improvements that decrease the energy we use to heat and cool our houses and apartments. More efficient practices include better insulation, passive solar heating, and the substitution of fans and open windows for air conditioning when practical. We can replace inefficient incandescent light bulbs with more efficient, compact fluorescent bulbs. A tremendous amount of energy is used in industrial manufacturing processes (see p.168), so there is also significant mitigation opportunity in simply being better about recycling.

There are other lifestyle changes we can make easily and immediately, ones that don't require that we remodel or even buy new bulbs or appliances. For example, in warm and dry weather, clotheslines make an excellent substitute for dryers. Appliances that are not in use can be unplugged, reducing electricity leakage.

We can make serious contributions to emission reduction efforts with our transportation choices. Many of us could commute to work by bicycle or on foot. For those of us who have difficulty finding time to maintain fitness regimes, this option allows for the best sort of multitasking—we exercise while we reduce our carbon footprints.

Other alternatives to driving long distances alone include public transportation and carpools. Fuel-efficient vehicles are another exciting new option. Given the high cost of gasoline in recent years, this option not only benefits the environment, but our pocketbooks as well.

Education and incentives

In addition to what individuals can do, employers, governments, and non-governmental organizations can play an important role. Community-focused organizations can provide relevant guidance and education to individuals. Some governments already provide tax benefits and incentives for citizens who build green, add solar panels to their roof, or buy hybrid vehicles. Public outreach efforts can also include educational programs that teach energy conservation practices, and campaigns aimed at encouraging individuals to make environmentally conscious decisions. If you want to know how well you are currently doing in terms of your own personal contribution to global greenhouse gas emissions, turn to p.182.

Decrease the amount of energy used in your home by installing simple solar panels.

Appliances that are not in use can be unplugged, reducing electricity leakage.

Clotheslines make an excellent substitute for dryers.

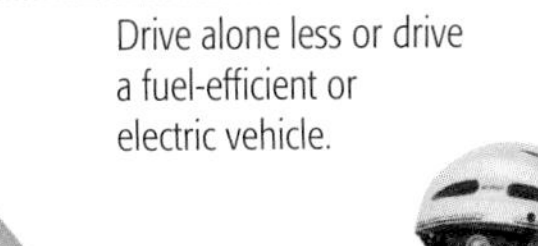

Drive alone less or drive a fuel-efficient or electric vehicle.

Remember to recycle.

Replace incandescent bulbs with fluorescent bulbs.

Commute to work by bicycle or on foot.

What's your carbon footprint?

Before you go on a diet, you might weigh yourself to establish a starting point and also, perhaps, to further motivate yourself to cut back on calories. In a similar way, you can calculate your "carbon footprint"—your personal contribution to the problem of global warming—as a first step in reducing the emissions your lifestyle generates.

There are numerous carbon footprint calculators on the Internet; we provide a few Web site addresses at the end of this discussion for your reference. All of these calculators help you to evaluate your lifestyle and determine your personal contribution or "footprint." Carbon footprints are usually measured in metric tons of CO_2 eq per year.

In order to assess your footprint, you will be asked to answer a series of questions about your lifestyle.

Footprint questions

- Where do you live? There are regional differences in the energy sources used to generate electrical power and these affect emissions.
- How many people live in your home? Your carbon footprint is smaller if the energy you use is shared.
- What type of vehicle and how many miles per year do you drive? Your car's gas mileage affects emissions.
- How often do you fly, and are your trips short or long? Airline travel is a big emissions contributor. Short trips use more energy per mile because takeoffs are particularly fuel-intensive.
- Do you heat your home with natural gas, heating oil, or propane? How much is your typical monthly heating bill? What is your typical monthly electric bill?
- Do you eat red meat, just chicken and fish, or are you a vegetarian? The various activities that together put these foods on your table have differing carbon emission profiles.
- Do you eat mostly local foods and buy mostly local products, or do you prefer imported goods? Long-haul freight consumes fossil fuel aggressively.
- What types of recreation do you prefer? A bicycling trip that begins at your own front door contrasts quite markedly with snowmobiling or motorboating at distant recreation sites.

Now take off your shoes and compare your print to two famous footprints!

SASQUATCH

Carbon footprint:

30 METRIC TONS CO_2 EQ PER YEAR

At home
Lives in a poorly insulated, air-conditioned apartment in the American Midwest with electric heat; loves to soak in a hot bathtub several times a week; does not recycle.

On the go
Drives an SUV 24,000 km per year; takes 5–10 business trips by plane each year, both short haul and long haul; and takes one exotic personal vacation each year by plane.

CINDERELLA

Carbon footprint:

4 METRIC TONS CO_2 EQ PER YEAR

At home
Lives with two roommates in France in an energy-efficient, well-insulated apartment with electric heat; prefers taking showers; recycles a significant fraction of household waste.

On the go
Drives a hybrid car 16,000 km per year; journeys by train 1600 km per year; uses videoconferencing rather than traveling for business; and takes one exotic personal vacation each year by plane.

Does your footprint looks more like Cinderella's dainty glass slipper or Sasquatch's big paw? Any of the following Web sites will help you get closer to the answer and suggest ways in which you can become more carbon stingy and environmentally friendly.

ON-LINE CARBON FOOTPRINT CALCULATORS

http://www.epa.gov/climatechange/emissions/ind_calculator.html
http://atmospheres.gsfc.nasa.gov/iglo/
http://www.carbonfootprint.com/
http://www.climatecrisis.net/takeaction/carboncalculator/
http://www.bp.com/extendedsectiongenericarticle.do?categoryId=9015627&contentId=7029058

Global problems require international cooperation

The atmosphere does not recognize national boundaries. When pollutants such as greenhouse gases or industrial aerosol particulates are emitted, they travel great distances, crossing continents and oceans. No single nation can solve the problems created by atmospheric pollutants. Some argue that, because of its pervasiveness, global warming is too daunting a challenge, that humans simply cannot solve a problem so huge in scale—especially one that requires cooperation between so many disparate parties. History, however, suggests otherwise, and even provides precedent for cooperation between nations in solving environmental problems.

Acid rain

By the 1970s, large parts of the northeastern United States and eastern Canada were plagued by the problem of "acid rain." As lakes, rivers, and ponds became acidified, fish populations and other aquatic life died off in startling numbers. Acidic rainfall began to kill trees as well. Research implicated the sulphate and nitrate aerosol particulates produced by factories located in the American Midwest. Aerosols were being carried downwind, where they dissolved in rainfall to form sulfuric and nitric acid. This so-called "acid rain" ultimately ended up in streams, rivers, ponds, and lakes. Europe also experienced similar acid rain problems. In some cases, the acid rain was destroying historic monuments and structures, as well as natural habitats.

In response to these and other related problems, the United States passsed a series of laws, culminating with the "Clean Air Act" of 1990. This act included specific provisions for dealing with the acid rain problem. The Clean Air Act of 1990 (and other clean air acts passed in the US and other nations) led to the widespread introduction of "scrubbers" into factories, which remove harmful particulates from industrial emissions before they enter the atmosphere.

The effects of acid rain
This conifer forest on Mount Mitchell, in North Carolina, includes many trees damaged or killed by acid rain.

Ozone depletion

Perhaps an even better analogy to the challenge posed by climate change is the problem of ozone depletion (see p.30). A breakdown in the stratospheric ozone layer was measured at the South Pole in the early 1980s. Theoretical and observational considerations pointed to an anthropogenic cause. Industrial products such as chlorofluorocarbons (CFCs), used at the time as refrigerants and propellants in aerosol cans, were eventually reaching the stratosphere. There, in the presence of solar radiation, they produced chemicals capable of destroying the ozone layer.

Since the ozone layer prevents most harmful ultraviolet radiation from reaching Earth's surface, the depletion of this protective layer was tied to an increase in skin cancers, and other damaging effects on plants and animals. In 1989, worldwide concern led to adoption of the Montreal Protocol, an international agreement banning the production of ozone-depleting chemicals. Former UN Secretary-General Kofi Annan has referred to the protocol as:

CFCs

CFCs were used as propellants in spray cans until they were implicated in the destruction of the ozone layer.

"Perhaps the single most successful international agreement to date."

Former UN Secretary General Kofi Annan

A global challenge

Arguably, the problem of climate change is more challenging to solve than that of ozone depletion. In that case, other commercial refrigerants and propellants were readily available as substitutes for ozone-depleting substances. By contrast, the emission of greenhouse gases results from the world's dependence on fossil fuels—its primary source of energy—and unfortunately a substitute for fossil fuels that can meet current (and future) world energy demands has yet to be found. Emission control and reduction requires a fundamental change in global energy policies.

The Kyoto Protocol

Recent efforts to achieve just that have begun:

- **1992:** The United Nations Framework Convention on Climate Change (UNFCCC), an international treaty, was formulated at the "Earth Summit" held in Rio De Janeiro, Brazil.
- **1997:** Five years later, an update to the UNFCCC—the " Kyoto Protocol"—was agreed upon at a summit in Japan.
- **2005:** Only after another eight years did the treaty actually go into force. The stated objective of the Kyoto Protocol is to achieve "stabilization of greenhouse gas concentrations in the atmosphere at a level that would prevent dangerous anthropogenic interference with the climate system."
- **At present:** 172 parties have currently ratified the Kyoto Protocol, including all but two major industrial countries. All of the parties that have entered into the agreement are committed to reducing greenhouse gas emissions to mandated levels. The two industrialized nations that have not ratified the protocol are the United States and Australia. However, the Australian Prime Minister elected in November of 2007 has pledged to sign. A large number (137) of developing nations have also ratified the protocol, but they are not held to mandated reductions in light of the financial hardships doing so might impose upon their fragile economies. The protocol includes provisions to insure that the developed world assists developing nations in moving toward more environmentally friendly energy resources.

The Kyoto Protocol has been criticized by both sides in the climate change debate. Critics on one side argue that it doesn't go far enough, and that the emission cuts mandated in the protocol will not stabilize greenhouse gas concentrations below dangerous levels. Supporters point out, however, that it is just a first step, putting in place a framework that can be built upon in the future to achieve broader reductions. Critics on the other side argue that committing to Kyoto will destroy the global economy. Yet cost-benefit analyses suggest that the cost of inaction could

Logging concessions
A timber truck transports logs in the Miri interior, eastern Malaysian Borneo state of Sarawak. A group of developed nations and an American green group donated U.S. $160 million for a World Bank-led climate-change plan in December 2007, which encourages developing nations to conserve their tropical forests. The project was launched in Bali, Indonesia, amid negotiations to establish a new global climate change agreement for 2012 and beyond.

be far greater (see p.146). The current debate is arguably more about politics than objective scientific or economic considerations; as such, it is likely to continue into the forseeable future.

Post-Kyoto period

Many feel that an additional breakthrough is still needed to produce an effective global agreement to stabilize greenhouse gas emissions. An obvious timeframe for this would be 2012, when the commitment period for the Kyoto Protocol expires. In December of 2007, further negotiations took place at a conference in Bali, Indonesia. The purpose of this conference was to establish a new global greenhouse emissions agreement for the post-Kyoto period, lay groundwork for the required international negotiations, and begin to establish a timeline for finalization of the new agreement. The various delegations, including major players such as the US, the European Union, China, Japan, and Canada, agreed to a "roadmap" in Bali for future negotiations. However, certain delegations (e.g., the US, Canada, and Japan) opposed specific cutback targets championed by the EU and others. As a result, the roadmap is a compromise between the competing interests of different nations. The UK Prime Minister, Gordon Brown, called the agreement "a vital step forward for the whole world." However, he warned that the agreement was "just the first step…now begins the hardest work, as all nations work towards a deal in Copenhagen in 2009 to address the defining challenge of our time."

Climate change summit
A new roadmap for addressing human-caused climate change was reached at the UN Summit in Bali in December 2007.

Can we achieve sustainable development?

A responsible society strives to meet its needs without compromising the ability of future generations to meet theirs. This defines sustainable development. Sustainability requires that we protect ecosystems from destruction and consume natural resources at a rate no greater than nature can provide. Achieving environmental sustainability is difficult in terms of water use and soil erosion, and seemingly impossible when we take into account fossil-fuel and mineral-resource consumption. Since cutting consumption is such a challenge, we should also look to renewable energy and recycling to achieve sustainability goals.

Developing nations

Issues of equity enter into the sustainability equation because current models of economic development depend on increased consumption and depletion of natural resources. The challenge is to find ways that developing countries can achieve a quality of life equal to that of the developed world without damaging the environment and depleting resources. Doing so requires that they make a substantial shift away from the highly consumptive and largely unsustainable path followed by the developed world. Fortunately, developing nations can now utilize previously unavailable technologies that may help them to meet their needs with reduced impact on the environment.

For example, China has slowed its fossil-fuel-use increases with a combination of activities. By shifting to renewable and less carbon-intensive energy sources, imposing economic reforms, and slowing population growth they have moved in a positive direction. India, Turkey, Mexico, South Africa, and Brazil are also working to decouple economic development from fossil-fuel dependency.

SUSTAINABLE DEVELOPMENT STRATEGIES

There are many strategies for mitigating against climate change. Most of these enhance sustainability but also involve trade-offs.

	Mitigation option		
	Improving energy efficiency	Reforestation	Deforestation avoidance
Compatibility with sustainable development	Cost effective; creates jobs; benefits human health and comfort; provides energy security	Slows soil erosion and water runoff	Sustains biodiversity and ecosystem function; creates potential for ecotourism
Trade-offs		Reduces land for agriculture	May result in loss of forest exploitation income and shift to wood substitutes that produce more emissions

In many cases, sustainable development strategies are clearly win-win, as is the case with alternative energy or carbon sequestration (see p.178). These technologies can enhance national security by reducing dependency on foreign oil, create new jobs, and stimulate economies (see table below). But in other instances, short-term economic benefits may conflict with environmental benefits. For example, the shift in developing countries from biomass (wood-fire) cooking to the use of cleaner and more efficient liquid propane (fossil-fuel) stoves enhances human health and quality of life by reducing indoor pollution, but increases dependency on fossil fuels. This, in turn, increases greenhouse gas emissions and exacerbates human-induced climate change.

Developing policy

The responsibility for implementing sustainable development policies lies with government, industry, and civil society:

- Just as human health has improved globally thanks to diverse local strategies, progress toward a common goal such as climate change mitigation can be made via disparate governmental policies.
- Sustainability and profitability are gradually being seen as compatible goals in industry, perhaps essentially so for large, multinational companies. Regulatory compliance is also a factor, but may not be as important as was once thought.
- Non-governmental organizations (NGOs), which encourage reform, provide policy research and advice, and champion environmental issues, are increasingly expressing the will of civil society. Academia also plays a role, especially in research, which enhances understanding of the scientific, economic, and political implications of climate change. The IPCC itself is a prime example of how a civil body can influence the world's response to global warming.

Fortunately, the goals of climate change mitigation and sustainable development are largely compatible. Together, these two strategies can help us to create a healthier and more durable society for the future.

Incineration of waste	Recycling	Switching from domestic fossil fuel to imported alternative energy	Switching from imported fossil fuels to domestic alternative energy
Energy is obtained from waste	Reduces need for raw materials; creates local jobs	Reduces local pollution; provides economic benefits for energy-exporters	Creates new local industries and employment; reduces emissions of pollutants; provides energy security
Air pollution prevention may be costly	May result in health concerns for those employed in waste recycling	Reduces energy security; worsens balance of trade for importers	Alternative energy sources can cause environmental damage and social disruption, e.g., hydroelectric dam construction

The ethics of climate change

The international media has paid considerable attention to the economic implications of global climate change. They have, by contrast, paid little attention to the equally important ethical considerations. The objective of the Kyoto Protocol itself—to stabilize "greenhouse gas concentrations…at a level that would prevent dangerous anthropogenic interference with the climate system"—begs several questions. Who, for example, determines what constitutes "dangerous"? Answering such questions requires us to take into account political, cultural, and philosophical principles that are fundamentally ethical in nature.

After the flood
A Bangladeshi woman collects water from a well submerged by flood water at Paikpara. Hundreds of people in the northern districts of the country sought shelter after their houses were lost in the severe flood of July 2007.

Winners and losers

One tricky ethical principle is "equity." Equity issues surrounding climate change include the fair distribution of risks, benefits, responsibilities, and costs to both developed and developing nations. Climate change will be associated with potentially dramatic redistributions of wealth and resources, impacting food production, fresh water availability, and environmental health. In the course of this shuffling, there will be winners and losers. Unfortunately, climate change won't play fair. In fact, climate change may play the role of a "reverse Robin Hood," taking resources from the poor, and giving them to the rich. Tropical regions—the developing world essentially—will likely suffer the most detrimental impacts. In the short term, the developed nations may even stand to benefit. Europe and North America agribusinesses, for example, may enjoy longer growing seasons (see p.130).

Further ethical complications arise from the fact that the individuals who gain from current fossil-fuel burning are not the same as the individuals who stand to lose when the climate changes. Is it possible to assign a meaningful cost to the devastating impacts of climate change on the poor and disadvantaged?

What is the value of the life of a starving child in Bangladesh as measured in cheap barrels of oil? Do we even dare to pose such questions?

And the developing world, by virtue of its relative poverty and lack of technological infrastructure, is far more vulnerable to the economic, environmental, and health threats posed by climate change. Ethical considerations would seem to demand that the developed world assist developing nations in adapting to climate change, both in mitigating impacts, and exploiting possible benefits.

The developed world has already benefited from a century of cheap fossil-fuel energy. Given this fact, it is surely unfair to tell developing nations, who are just now beginning to build their energy and transportation infrastructures, that they can't have their turn to enjoy cheap oil. This challenging ethical dilemma complicates discussions about the appropriate burden of mitigation, including the distribution of emissions rights both among nations and between generations.

When it comes to the generational transfer of the benefits and costs of fossil-fuel burning, "social discounting" places a greater value on benefits today at the expense of subsequent generations. This is based on the assumption that future generations will have access to new technology and will be better equipped to deal with environmental challenges. Social discounting involves making an ethical call. If we discount future potential impacts too strongly and assume that future generations will be able to solve all problems, the cost-benefit analysis will surely favor inaction. Is it fair to gamble like this, knowing that it is our grandchildren who will pay the price if our assumptions turn out to be wrong?

Geoengineering dilemmas

Ethical considerations are also raised by geoengineering approaches to mitigation (see p.178). Should nations that stand to benefit from certain types of interference be able to do so, even when other nations may be negatively impacted by their actions? Ethical issues complicate discussions of biofuel technology too. Should agricultural land currently used to feed people be reallocated for energy production at a time when starvation and malnourishment are omnipresent?

An essential step in tackling these problems is for all countries, including holdouts like the United States, to join the international effort to stem the buildup of greenhouse gases initiated by the Kyoto Protocol (see p.186).

The known unknowns and the unknown unknowns

There are at least two kinds of unknowns. There are the "known unknowns," which are the questions we already know to ask, but for which we don't yet have the answers. Then there are the "unknowns unknowns." These are the questions we don't even know to ask, the questions involving phenomena that currently lie beyond the horizons of our imagination.

A great deal of discussion in this book is devoted to the known unknowns. We have discussed the open scientific questions regarding how much warming is to be expected and precisely what the pattern of climate change will be. These uncertainties are linked to unknowns regarding the societal and environmental impacts of climate change (e.g., changes in water availability, food supply, and disease prevalence). We have examined the still unsolved mysteries of the great climate changes in Earth's past, and the changes in violent weather phenomena, such as hurricanes, that may lie in store for us in the future.

More known unknowns
The known unknowns also include the lack of certainty regarding the "tipping points" looming in our future. Scientists recognize that such tipping points probably exist, but they don't know exactly where they may lie:

- Just how rapidly will the major ice sheets melt, and how high will the sea level rise accordingly?
- Will the "conveyor belt" ocean circulation weaken? And if so, when?
- Will the ability of the oceans and plants to absorb the CO_2 we are adding to the atmosphere change in the future?

Also included in the known unknown category are answers to questions relating to the unpredictability of human behavior.

- What will future human-driven emissions patterns be?
- What will the economic implications of warming be?
- What steps will we take to mitigate against greenhouse gas buildup and climate change? How successful will mitigation efforts be?
- Will we implement any of the currently conceived geoengineering plans? Will new, risk-free plans be conceived of?

Unknown unknowns

And what about the unknown unknowns? There are some of these in the science itself:

- Will the response of the climate to increased greenhouse gas concentrations take an unpredicted course?
- What are the tipping points that have not been conceived of yet?
- Are there hidden reserves of carbon on our planet that could suddenly be released, leading to further warming?

In the case of adaptation and mitigation, the unknown unknowns may be the stuff of science fiction. Decades ago, who would have imagined modern-day technology such as cloning or hand-held "smart phones" as powerful as the supercomputers of previous decades? More to the point, who would have conceived of modern transportation options such as hybrid vehicles, or prospective energy technology such as "cellulosic ethanol" (see p.172)?

So what are we to make of all of this uncertainty?

Clearly, we must work to diminish the uncertainty where possible, particularly when it impacts on our ability to make appropriate policy decisions or choose an optimal strategy for mitigating climate change. Recent history has taught us that uncertainties are not adequate justification for avoiding action. We know enough today to understand how vital it is that we act now.

A scuba diver explores a dark underwater cave
What surprises are in store for us as we continue to probe the climate system's sensitivity to human insult?

The urgency of climate change
Why we must act now

Uncertainty abounds (see p.192) but it is a poor excuse for inaction. In fact, given the possibility of severe and irreversible harm to society and the environment, scientists generally advocate that we abide by a "precautionary principle" that puts the onus of proof on those advocating inaction.

No excuses

If the remaining uncertainty in the science is not a valid argument against taking immediate action to slow climate change, then what, if anything, is? As we have seen, some argue that action could harm the economy. Yet this argument does not appear to withstand scrutiny, since the economic harm of inaction looks to be greater (see p.146) in the long term. Others argue that climate change might be beneficial to humankind, but an impartial assessment strongly suggests otherwise (see p.104). Still others concede that climate change represents a potential threat, but that it is only one of many problems facing society, and that focusing on climate change issues might divert attention and resources from more pressing problems. The argument that we must choose between competing societal problems, however, is based on the flawed premises that society can only solve one problem at any given time, and that the problems facing society are independent of each other. In the case of climate change, we have already shown its potential to exacerbate other major global societal and environmental issues, including:

- Sustainability
- Regional conflict
- Biodiversity
- Extreme weather events
- Water availability
- Disease

Some proponents of inaction argue that we can engineer our way out of the problem with future technological "fixes." However, the potential pitfalls of high-tech fixes present risks as well. And many of these fixes may not be able to prevent or reverse the more serious consequences of climate change, such as the melting of the Greenland ice sheet. While there are some promising new technologies on the horizon, there are none currently available that will handily satisfy our global thirst for carbon-free or carbon-neutral energy in the decades to come. Furthermore, while carbon sequestration (CCS) may ultimately slow the buildup of carbon dioxide in the atmosphere, the feasibility of large-scale implementation of CCS has yet to be demonstrated (see p.178).

Climate change has been described as a problem with a huge "procrastination penalty." With each passing year of inaction, stabilizing Earth's climate becomes increasingly difficult.

A polar bear sits on a small iceberg
Will these amazing creatures be an early casualty of human-caused climate change?

Our children's world

If we choose not to act on this problem now, then in the very best-case scenario we must accept that our children and grandchildren will grow up in a world lacking some of the beauty and wonder of our world. They may come of age in a time where:

- Polar bears, golden toads, and numerous other creatures will be the stuff of myth
- There will be no Great Barrier Reef to explore
- Giraffes and elephants will no longer loom in the foreground of the majestic snows of Kilimanjaro
- Great coastal communities such as Amsterdam, Venice, and New Orleans will join the lost city of Pompeii

Of course, humankind might plausibly adapt to these sad changes.

In the worst case scenario, however, our grandchildren will grow up, as renowned climate scientist James Hansen has bluntly put it, on "a different planet"—a planet potentially resembling the dystopian world depicted in science fiction movies such as *Soylent Green* and *The Island*. Adaptation, in this case, is unlikely to be viable for many of the world's people and other living things.

Where does that leave us?

There is no "silver bullet" that will solve the problem of global climate change. But that does not mean we should throw up our hands in the face of this urgent problem. Any viable solution is going to require action from many governments and all strata of society; it will involve adapting to the changes that are inevitable, and mitigating the changes we can avert. It goes without saying that alternative energy sources must be aggressively developed and deployed, and that governments must incentivize and reward responsible behavior by individuals and corporations.

The future in our hands
Our planet has supported life for billions of years, but only over the past century has a species—humans—developed the ability to alter the planetary environment. Will we do good or harm with this newfound ability? The answer is in our hands.

Climate change is one of the greatest, if not **the greatest** challenge ever faced by human society.

But it is a challenge that we must confront, for the alternative is a future that is unpalatable, and potentially unlivable. While it is quite clear that inaction will have dire consequences, it is likewise certain that a concerted effort on the part of humanity to act in its own best interests has great potential to end in success.

Glossary

Acid rain

Acid rain refers to any form of precipitation (e.g., rain, snow, sleet) that is unusually acidic. Largely caused by industrial emissions of sulfur and nitrogen aerosols, which form sulfuric and nitric acid when combined with water droplets suspended in the atmosphere, acid rain has caused documented damage to trees and plants, fish and other aquatic animals, building facades, and monuments. In recent decades, the governments of the US, the UK, and Canada have passed legislation, such as the Clean Air Acts, to reduce the industrial emissions responsible for acid rain.

Aerosols

Aerosols are microscopic liquid droplets, dust, or particulate matter that are airborne in the atmosphere. An aerosol may remain suspended in the atmosphere for hours or years depending on the type of aerosol and its location in the atmosphere. Aerosols can be of either human or natural origin and they may reflect and/or absorb incoming and outgoing radiation. Consequently, aerosols impact the atmospheric energy budget and temperatures within Earth's atmosphere and at Earth's surface. This scientific definition of aerosol should not be confused with the common usage in association with so-called "aerosol" spray cans and their contents. (See glossary entry for Chlorofluorocarbons.)

Atmosphere

The atmosphere is the gaseous envelope surrounding Earth, which is retained by Earth's gravitational pull. The first 80 km above Earth contains 99% of the total mass of Earth's atmosphere and is generally of a uniform composition (except for a high concentration of ozone, known as the stratospheric ozone layer, at 19 to 50 km). The gases that make up the atmosphere are nitrogen, 78.09%; oxygen, 20.95%; argon, 0.93%; carbon dioxide, 0.04%; and minute traces of neon, helium, methane, krypton, hydrogen, xenon, and ozone as well as trace amounts of water vapor, the distribution of which is highly variable. Earth's atmosphere features distinct layers: the troposphere, the stratosphere, the thermosphere, and the exosphere. The term "free atmosphere" refers to the portions of the atmosphere that lie above the troposphere. (See glossary entries for Stratosphere and Troposphere.)

Biofuels

Biofuels are solid, liquid, or gaseous fuels consisting of, or derived from biomass (plant material, vegetation, or agricultural waste used as energy sources). Biofuels can aid in the mitigation of greenhouse gas emissions by providing carbon neutral alternatives to fossil fuel burning. Burning biofuels does not lead to any net increase in greenhouse gas concentrations because the carbon released when biofuels burn was just recently removed from the atmosphere and has only been stored in plants for a few months. For every gram of CO_2 released when a biofuel is burned, a gram was removed from the atmosphere by photosynthesis just a short time ago. This balance is why biofuels are considered "carbon neutral." Liquid or gaseous biofuels can be used for transport, while solid biofuels can be burned for heat or to generate electric power. Common currently used biofuels, such as ethanol, are derived from maize (corn). Cleaner and more efficient biofuels, such as cellulosic ethanol, derived from switchgrass, are currently under development.

Carbon cycle

The carbon cycle is the sum of the processes, including photosynthesis, decomposition, respiration, weathering, and sedimentation, by which carbon cycles between its major reservoirs: the atmosphere, oceans, living organisms, sediments, and rocks. (See glossary entry for Photosynthesis.)

Carbon dioxide (CO_2) equivalent

Carbon dioxide equivalent expresses the amount of CO_2 that would have the same global warming potential as a given greenhouse gas measured over some defined timeframe, typically one century. It is typically measured in gigatons ("Gt CO_2 eq"). (See glossary entry for Gigaton.)

Chlorofluorocarbons (CFCs)

Chlorofluorocarbons (CFCs) are synthetic compounds consisting of a carbon atom surrounded by some combination of chlorine and fluorine atoms. CFCs are powerful greenhouse gases. However, they are better known for their role in ozone depletion. CFCs were formerly widely used as refrigerants, propellants (in so-called "aerosol" cans), and as cleaning solvents. When their use was implicated in the destruction of the protective stratospheric ozone layer in 1989, the industrial use of CFCs, and other related compounds, was prohibited by the Montreal Protocol. (See glossary entry for Ozone.)

Climate forcing

Changes in the global energy or "radiative" balance between incoming energy from the Sun and outgoing heat from Earth lead to changes in climate. There are a number of mechanisms that can upset this balance, for example fluctuations in Earth's orbit, and changes in the composition of Earth's atmosphere. In recent times, Earth's atmospheric composition has changed as a result of human-generated greenhouse gas emissions. By altering the global energy balance, such mechanisms "force" the climate to change. Consequently, scientists call them climate forcing or radiative forcing mechanisms.

Climate proxy
Climate proxies are indirect sources of climate information from natural archives such as tree rings, ice cores, corals, cave deposits, lake and ocean sediments, tree pollen, and historical records. Information from climate proxies can be used to reconstruct climate for times prior to the establishment of a widespread instrumental atmospheric and oceanic data set. (See glossary entry for Instrumental record).

Cryosphere
The cryosphere is a term for the cold regions of the planet where water persists in its frozen form, i.e., regions covered with glaciers and ice sheets, or with permanently frozen soils. The cryosphere plays different roles within the climate system. The two continental ice sheets of Antarctica and Greenland actively influence the global climate and may also have effects on sea level. Snow and sea ice, with their large areas and relatively small volumes, are connected to key interactions and feedbacks on global scales, including solar reflectivity and ocean circulation. Perennially frozen ground (permafrost) influences soil water content and vegetation over vast regions, and is one of the cryosphere components that is most sensitive to atmospheric warming trends.

Deep time
Distant geologic history, generally before the recent ice-age cycles began 2 million years ago, is often referred to as deep time.

El Niño
El Niño is a climate event in the tropical Pacific ocean and atmosphere wherein the trade winds in the eastern and central tropical Pacific are weaker than usual, there is less upwelling of cold subsurface ocean water in the eastern Pacific, and relatively warm water spreads out over much of the tropical Pacific ocean surface. During an El Niño event, the warmer tropical Pacific surface ocean waters influence the overlying atmosphere and alter the patterns of the extratropical jet streams of the northern and southern hemisphere and the general circulation of the atmosphere. The altered circulation of the atmosphere leads to changes in temperature and precipitation patterns in many regions across the globe. The name "El Niño" (literally "the boy child") derives from the Spanish term for the Christ Child and originates in the fact that the warming of the ocean waters off the Pacific coast of South America is usually most pronounced around Christmas time. (See glossary entries for ENSO, Jet stream, and La Niña.)

El Niño/Southern Oscillation (ENSO)
The El Niño/Southern Oscillation or ENSO phenomenon is an irregular oscillation in the climate involving interrelated changes in ocean surface temperatures and winds across the equatorial Pacific, which influences seasonal weather patterns around the world. ENSO is associated with alternations between El Niño climate events in certain years and La Niña events in others. (See glossary entries for El Niño and La Niña.)

Energy balance model (EBM)
An energy balance model (EBM) is the simplest type of global climate model. An EBM can be used to determine average global temperature by computing the balance between incoming (solar) and outgoing (terrestrial) radiation. (See glossary entry for General circulation model.)

Fossil fuels
Fossil fuels are hydrocarbon-based energy sources formed over millions of years when the fossilized remains of dead plants and animals are exposed to heat and high pressure in Earth's crust. Fossil fuels exist in solid form as coal, shales, and methane "clathrate" (methane gas entrapped in a water-ice cage); in liquid form as oil; and in gaseous form as so-called natural gas (mostly methane). Nearly 90% of the world's primary energy production comes from the combustion of fossil fuels. When fossil fuels are burned, greenhouse gases are released into the atmosphere. (See glossary entry for Greenhouse gases.)

Fuel cell technology
A fuel cell can be used to convert fuel energy into electricity in a manner similar to, but distinct from, a battery. In a fuel cell, electricity is generated from the reaction of chemical fuel stored at one end of the cell with an oxidant at the other end of the cell. Unlike in a conventional battery, reactants are consumed during the operation of the fuel cell and must therefore be replaced for continuous electricity production. The primary application of fuel cell technology is in the area of transportation. There are a variety of fuels that can potentially be used in fuel cells. If hydrocarbon-based fuels are used, fuel cells emit only marginally less greenhouse gases than conventional carbon-based energy sources. However, alternatives such as hydrogen cell technology, which are currently being researched, could provide a carbon-free alternative to fossil-fuel based transportation.

General circulation model (GCM)
A general circulation model (GCM) is a three-dimensional numerical model used in global climate prediction and assessment. Unlike simpler energy balance models (EBMs), GCMs can be used to solve for a variety of variables including wind patterns, air pressure, atmospheric humidity, and precipitation patterns. While the most basic GCMs model the behavior of the atmosphere alone, climate modelers often use

"coupled" versions of GCMs wherein the atmosphere is allowed to interact with models of the global oceans, the major (Greenland and Antarctic) ice sheets, and terrestrial ecosystems. (See glossary entry for Energy balance model.)

Geothermal energy

Geothermal energy is generated from the heat stored beneath Earth's surface, typically through the use of steam-driven turbines. Often water is injected into the hot subsurface of Earth to generate steam. The use of geothermal energy sources dates back to the early 20th century. Geothermal power provides less than 1% of global energy production, but it is used more consistently in certain regions such as Iceland, and California and Nevada in the United States. Potential exists for more widespread use of this renewable energy source.

Gigaton (Gt)

A gigaton (Gt) is a metric unit of mass, equal to 1 billion metric tons (tonnes). In the context of greenhouse gas emissions, gigatons are commonly used as units for measuring global quantities of carbon dioxide or carbon. (See glossary entry for Carbon dioxide (CO_2) equivalent.)

Global warming potential (GWP)

Global warming potential (GWP) is a measure of how much a given mass of greenhouse gas is estimated to contribute to global warming relative to the same amount of carbon dioxide (see p.28). Since GWP is a measurement of the integrated warming impact of greenhouse gas emissions, it must be calculated and stated for a specific time interval, typically one century.

Greenhouse gases (GHGs)

Greenhouse gases (GHGs) are gases in Earth's atmosphere that absorb longwave radiation including the radiation emitted from Earth's surface (i.e., terrestrial radiation). Because they absorb terrestrial radiation, these gases have a warming influence on Earth's surface (referred to as the "Greenhouse Effect"). Greenhouse gases exist naturally in Earth's atmosphere in the form of water vapor, carbon dioxide, methane, and other trace gases, but atmospheric concentrations of some greenhouse gases such as carbon dioxide and methane are being increased by human activity. This occurs primarily as a result of the burning of fossil fuels, but also through deforestation and agricultural practices. Certain greenhouse gases, such as the CFCs, and the surface ozone found in smog (which is distinct from the natural ozone found in the lower stratosphere), are produced exclusively by human activity.

Hadley circulation/hadley cell

The pattern of rising moist air near the equator and sinking dry air in the subtropics is referred to as the "Hadley Cell" or the "Hadley Circulation" after the 18th-century English amateur scientist George Hadley who first formulated a theory about this atmospheric circulation system. The Hadley Circulation is a key component of the general circulation of the atmosphere; it helps to transport heat from the equatorial region to higher latitudes and is responsible for the trade winds (easterly surface winds) in the tropics (see p.89).

Hydropower

Hydropower is power produced by capturing the kinetic energy of moving water. It is currently the most commonly used source of renewable energy, responsible for just over 6% of global energy production. Hydropower has been used in primitive forms (e.g., for powering gristmills or for irrigation) for many centuries. While obtaining direct mechnical energy from hydropower requires proximity to a moving water source, modern conversion of hydropower to electric power allows long-distance transport of energy, albeit with energy loss (that increases with distance from the source). The availability of hydropower on a regional basis in the future could be affected by shifting patterns of rainfall and runoff associated with climate change.

Ice sheets/Ice Age/glaciers/glaciation/glacial/interglacial

Glaciers are huge masses of ice formed from compacted snow. An ice sheet is a mass of glacier ice that covers surrounding terrain and is greater than 50,000 km^2. (The only current ice sheets are in Antarctica and Greenland.) An ice age is a cold period resulting in an expansion of ice sheets and glaciers. This expansion is referred to as glaciation. Ice ages are marked by episodes of extensive glaciation alternating with episodes of relative warmth. The colder periods are called glacials, the warmer periods are referred to as interglacials.

Instrumental record

In the context of climate data, the instrumental record refers to the relatively brief record of direct measurements recorded by instruments such as thermometers, barometers, rainfall gauges, and other devices that measure atmospheric temperature, pressure, wind, humidity, and precipitation, as well as ocean temperature, salinity, water density, and currents. For the atmosphere and ocean surface only, widespread measurements are available as far back as 100 to 150 years. For the free atmosphere and deep ocean, such measurements are generally only available for the past five or six decades. A few isolated instrumental climate records from several centuries back in time are available for regions such as Europe. To obtain climate data from the distant past, climate scientists turn to climate proxy records. (See glossary entry for Climate proxy.)

Intertropical convergence zone (ITCZ)

The Intertropical Convergence Zone (ITCZ) is a belt of low surface pressure that is centered near the equator but migrates north and south within the tropics as the seasons change. The ITCZ is associated with trade winds that converge near the equator, ascending as warm, moisture-laden air currents that rise deep into the upper troposphere in towering cumulous clouds and rainfall-producing thunderstorms. These winds eventually sink in subtropical latitudes but not before their original high moisture content has dissipated. (See glossary entry for Hadley circulation.)

Isotope

Isotopes are atoms of the same element having the same atomic number but different mass numbers. The nuclei of isotopes contain identical numbers of protons but have differing numbers of neutrons. Isotopes of a given element have the same chemical properties but somewhat different physical properties. Some isotopes are radioactive, which makes them useful for dating ancient materials (e.g., carbonaceous materials, rocks, etc.).

Jet stream

The jet stream is a high-speed wind current that lies roughly at the boundary between the troposphere and stratosphere at 8–17 km above Earth's surface. The major jet stream of each hemisphere (referred to as the "polar jet stream") is located at middle/sub-polar latitudes, while a weaker "subtropical jet stream" is found at lower, sub-tropical latitudes in each hemisphere. Both of the jet streams circle the globe as westerly winds (i.e., winds moving from west to east), and, like the ITCZ shift north and south with the seasons. (See glossary entries for Troposphere, Stratosphere, and ITCZ.)

La Niña

In a La Niña event, the trade winds in the eastern and central tropical Pacific are stronger than usual, there is greater upwelling of relatively cold subsurface ocean water in the eastern Pacific, and that cold water spreads out over tropical Pacific ocean surface. During a La Niña event the tropical Pacific ocean and atmosphere are in the opposite state as they are during an El Niño event and the influence on atmospheric circulation and global weather patterns is roughly, though not precisely, the opposite. The term "La Niña" means "the girl child" in Spanish. (See glossary entries for El Niño and ENSO.)

Longwave radiation

Longwave radiation (sometimes referred to as "infrared" radiation) is electromagnetic radiation typically associated with heat or thermal radiation. Atmospheric infrared radiation is monitored to detect trends in the energy exchange between Earth and its atmosphere. These trends provide information on long-term climate changes. Along with solar radiation, longwave radiation is one of the key components of Earth's energy balance studied by climate researchers. (See glossary entry for Solar radiation.)

Megaton (Mt)

A megaton (Mt) is a metric unit of mass, equal to 1 million metric tons (tonnes). There are 1000 megatons in a gigaton. In the context of greenhouse gas emissions, megatons are commonly used as units for measuring global quantities of carbon dioxide or carbon. This usage should not be confused with the distinct usage of the same term to describe the explosive power of a nuclear weapon.

Metric ton

One metric ton (tonne) is 1000 kilograms, roughly equivalent in mass to the Imperial ton (2200 pounds).

Microwave

A microwave is a high-frequency electromagnetic wave; its wavelength is between infrared and short-wave radio wavelengths. Microwaves measurements made by satellites provide one means of monitoring atmospheric temperature changes.

North Atlantic Oscillation (NAO)

The North Atlantic Oscillation (NAO) is a measure of the strength and direction of the predominantly westerly winds that blow across the North Atlantic ocean. The measurement is based on the surface pressure difference between the subpolar and subtropical regions over the North Atlantic ocean. The size of this pressure difference, and the strength and direction of the surface winds, varies from year to year, as first noted by Sir Gilbert Walker in 1932. The NAO can have a profound influence on temperature and precipitation patterns of the North Atlantic and neighboring regions of Europe and North America, particularly during winter in the northern hemisphere.

Organization for Economic Co-operation and Development (OECD)

The Organization for Economic Co-operation and Development (OECD) is an international organization consisting of 30 developed countries that subscribe to a set of Western political and economic principles. The OECD was formed in 1948. It was originally a group of European nations empowered to administer the Marshall Plan for post-World War II European reconstruction. Later, the group was expanded to include non-European Western nations. In 1961, the group formed the Organization for Economic Co-operation and Development.

Ozone
Ozone is a compound of oxygen that is made up of three atoms (the oxygen gas we breathe contains two oxygen atoms). Ozone is an irritating gas, when encountered in surface air pollution. However in the stratosphere, a natural layer of ozone protects life on Earth from harmful ultraviolet radiation from the Sun. CFCs have been implicated in the destruction of this ozone layer. Scientists predict that global warming will lead to a thinner ozone layer, because as the surface temperature rises, the stratosphere will get colder, making natural replenishment of ozone slower and its destruction faster. (See glossary entries for Solar radiation and Chlorofluorocarbons.)

Paleoclimatology
Paleoclimatology is the study of Earth's climate system in prehistoric periods (the paleoclimate), covering the past few centuries to billions of years. Among the methods employed in paleoclimate studies are the simulation of past climates with different continental configurations, ice sheet distributions, and greenhouse gas concentrations using climate models such as EBMs and GCMs, and the reconstruction of past climates from empirical data such as proxy climate records. (See glossary entries for EBM, GCM, and Climate proxy.)

Photosynthesis
Photosynthesis is the process by which green plants and certain other organisms synthesize carbohydrates from carbon dioxide and water using light as an energy source. The common form of photosynthesis releases oxygen as a byproduct. Photosynthesis is part of the carbon cycle and is instrumental in removing CO_2 from the atmosphere. (See glossary entry for Carbon cycle.)

Proxy
See Climate proxy.

Salinity
The technical term for saltiness in water is salinity. Salinity influences the types of organisms that live in a body of water and the kinds of plants that will grow on land fed by groundwater. Salt is difficult to remove from water, so salt content is an important factor in human water use. The salinity differences between different water masses in the ocean is also a key determinant of large-scale ocean currents.

Solar radiation
Solar radiation is the radiation emitted by the Sun, which generates energy from the nuclear fusion reactions in its interior. Most solar radiation is in the form of visible light and some is in the form of ultraviolet and other wavelengths of radiation. (See glossary entry for Ozone.)

Stratosphere
The stratosphere is the layer of Earth's atmosphere that lies above the troposphere, extending from roughly 8–17 km above the surface (lower near the poles and higher near the equator) to about 50 km above the surface. The lower stratosphere contains a natural ozone layer, which absorbs ultraviolet solar radiation, warming the surrounding atmosphere. The warming of the atmosphere with height inhibits vertical air currents, making the stratosphere a highly stable regime of the atmosphere, in contrast to the troposphere that lies below it. (See glossary entry for Ozone.)

Trade winds
Trade winds are the easterly winds (i.e., winds that move from east to west) that are found near Earth's surface in tropical regions. The rising atmospheric currents found within the ITCZ are associated with the convergence of trade winds. The El Niño/Southern Oscillation (ENSO) is associated with a periodic alternation between weakening and strengthening trade winds in the eastern and central tropical Pacific. (See glossary entries for ITCZ and ENSO.)

Troposphere/free troposphere
The troposphere is the lowest layer of Earth's atmosphere, extending from the surface of our planet to between 8 and 17 km (lower near the poles and higher near the equator). The troposphere is sometimes subdivided into a planetary boundary layer, where the atmosphere is in contact with Earth's surface, and the free troposphere, defined as the layer of the troposphere above the planetary boundary layer. The boundary between the troposphere and the stratosphere above it is called the tropopause. The troposphere contains roughly three quarters of Earth's atmosphere by mass. What we normally think of as "weather" takes place almost exclusively within the troposphere. (See glossary entry for Atmosphere.)

Wind shear
Wind shear is a difference in wind speed and direction between slightly different altitudes. Wind shear, among other things, can determine whether or not conditions are favorable for tropical cyclone development.

Index

Note: References in **bold** refer to the Glossary.

I

J

K

L

M

N

O

Picture Credits

The publisher would like to thank the following for their kind permission to reproduce their photographs:

(Key: a-above; b-below/bottom; c-centre; f-far; l-left; r-right; t-top)

AIMS - Australian Institute of Marine Science: 115. **Alamy Images:** Acbag 158-159; Alaska Stock LLC 1; David Ball 176-177; Peter Bowater 156-157; Jason Bye 163; Ashley Cooper 181cr; Darren Core 148-149; Gary Crabbe 84-85; David R. Frazier Photolibrary, Inc. 112-113; David Sanger Photography 128-129; Andrew Fox 136-137; Dennis Frates 100; Esa Hiltula 28-29; Hoberman Collection UK 149tc; D. Hurst 181cb, 181clb; Jeff Morgan Hay on Wye 120; Darrin Jenkins 160-161; Andre Jenny 181fcr; Huw Jones 72-73, 156bl; Scott Kemper 172-173; Lancashire Images 149; Patrick Lynch 180br; Malcolm Park Wine and Vineyards 46-47; Ron Niebrugge 134-135; NOAA / Michael Dwyer 79bl; Wolfgang Pölzer 114; Robert Harding Picture Library Ltd 124tl; Joern Sackermann 149tl; Sami Sarkis Underwater 192-193; Steve Smith / SuperStock 150-151; Robert Stainforth 55; Ariadne Van Zandbergen 58-59; Visions of America, LLC / Joe Sohm 164-165, 184c; David Wells 170-171; Janusz Wrobel 181tr. **Byrd Polar Research Center - OSU (The Ohio State University, USA):** 58clb; Lonnie G. Thompson 58bl. **Corbis:** 70-71; Abir Abdullah / epa 190-191; Alan Schein Photography 157fbr; Bill Barksdale 157bl; Chip East / Reuters 185br; Alejandro Ernesto / epa 103; Jose Fuste Raga 156fbr; Annie Griffiths Belt 110-111; Gunter Marx Photography 5cl, 108-109; Dewitt Jones 101; Everett Kennedy Brown / epa 157br; NASA 30-31; Rafiqur Rahman / Reuters 125crb; Dick Reed 146bc, 146bl, 146br, 146-147, 147bc, 147bl, 147br; Jim Reed 6-7; Reuters / Beawiharta 90-91; Rickey Rogers / Reuters 123; Alexander Ruesche / epa 156br; Mike Theiss / Ultimate Chase 102; David Turnley 124bl, 130-131; Barbara Walton / epa 26-27; Nik Wheeler 126-127. **Digital Railroad:** Daniel Beltra / Greenpeace 2-3. **DK Images:** David Peart 116-117; Royal Museum of Scotland, Edinburgh 42tl. **Florida Keys News Bureau:** Andy Newman 62-63. **Michael and Patricia Fogden:** 119tl. **The Galileo Project:** 80bl. **Getty Images:** AFP 186-187; The Image Bank / Joanna McCarthy 194-195; David McNew 49b; National Geographic / Paul Nicklen 118b; Jewel Samad / AFP 187br; Science Faction / David Scharf 97tr; Stone / Ernst Haas 124cla; Stone / Frank Oberle 122. Stone / Lester Lefkowitz 144-145. **Greenpeace:** Ardiles Rante 174-175. **IODP / TAMU:** 40-41. **John Gunn, Earth & Space Research:** 92-93. **John MacNeill Illustration:** 178-179. **Hans Kerp, Forschungsstelle fuer Palaeobotanik, Westfaelische Wilhelms-Universitaet Muenster:** 42cl. **Prof. em. Bruno Messerli / Geographisches Institut - Physische Geographie - Universität Bern:** 58crb. **Moviestore Collection:** 44. **NASA:** 66. NOAA: 56-57. **PA Photos:** AP Photo / Bullit Marquez 18; AP Photo / Karel Prinsloo 58br. **PhotoEdit:** 125tl. **Photolibrary:** Ifa-Bilderteam Gmbh 75; Phototake Inc. / MicroScan MicroScan 42-43. **PunchStock:** Corbis 157fbl; Digital Vision/Peter Adams 32-33 (b/g); Photodisc / Doug Menuez 4cra, 19. RubberBall 181tc. **Science Photo Library:** British Antarctic Survey 32cr, 32tr; Bernhard Edmaier 14bl; Will & Deni Mcintyre 184bl; Susumu Nishinaga 95br; Friedrich Saurer 23bl, 23cl, 23tl, 72tl; SOHO / ESA / NASA 80-81. **Shutterstock:** Tyler Olson 5fcra, 17ca, 23br, 23cr, 23tr, 39fcr, 74cb, 74clb, 74crb, 75clb, 77ca, 107ca, 141, 155, 196-197. **Statoil:** Bitmap / Kim Laland for StatoilHydro 168-169. **Still Pictures:** Roger Braithwaite 5cla, 98-99. **Sun-Sentinel:** South Florida Sun-Sentinel 5ftr, 142-143. Jeffrey Totaro / Esto: 166, 167. **USGS:** 69t, 116tl, 116tr, 138-139. **Werner Berner:** 32crb.

Jacket images: Front: Alamy Images: Panorama Media (Beijing) Ltd. clb; Getty Images: Riser / Jeremy Walker t.

Author Acknowledgements

We would like to thank our colleagues who provided helpful comments on the book at its various stages, including Richard Alley, Klaus Keller, and Jean-Pascal van Ypersele. We would also like to thank our colleagues at Penn State, who have helped to foster a stimulating environment for discourse on the important topic of climate change and its impacts. These include, among many others, Richard Alley, Mike Arthur, Eric Barron, Tim Bralower, Sue Brantley, Bill Brune, Rob Crane, Ken Davis, Bill Easterling, Jenni Evans, Bill Frank, Kate Freeman, Sukyoung Lee, David Pollard, Jim Kasting, Klaus Keller, Ray Najjar, Art Small, Petra Tschakert, Anne Thompson, and Nancy Tuana. Michael Mann would like to acknowledge his colleagues at RealClimate.org for the many keen insights into the science of climate change that they have shared over the years. Lee Kump similarly acknowledges his fellow members of the Canadian Institute for Advanced Research.

The book benefited greatly from the superb editing skills and unflappable patience of Erin Mulligan at Prentice Hall, and the graphics and design genius of the Dorling Kindersley publishing professionals: Stuart Jackman (for the design and development of the book); Richard Czapnik (for the page layouts); and Sophie Mitchell (for keeping the process on track). We would also like to thank Clive Savage, David McDonald, and Johnny Pau for additional design work; Sue Malyan and Jenny Finch for additional editorial work; and Sue Lightfoot for the index.

Michael Mann dedicates this book to the memory of his brother, Jonathan, and to his daughter, Megan, who will grow up on an Earth whose destiny rests in our hands. He thanks his wife, Lorraine Santy, for her support, and his parents, Larry and Paula Mann for the encouragement they have always provided. Lee Kump thanks his wife, Michelle, and his children, Katie and Sean, for their patience and support during the writing of this book. He dedicates the book to his mother, Patty Kump, for planting the seeds of wonder, and his father, Warren Kump, for his apprenticeship into the world of science.